Pharmaceutical Positioning - Deep Positioning

How Molecules become Great Medicines

Written by Mike Rea

Illustrations by Matt Stone
Cartoons by Ricardo Galvão

Published by IDEA Pharma
www.ideapharma.com
© 2023 IDEA Pharma

Contents

Prologue 7

Introduction: Why 'Deep Positioning' 9
    Summary 14
    Who is this book for? 15
    Principles of Deep Positioning 17

What is Deep Positioning? 18
    Ending up somewhere different 18
        Summary 21
    The 'North Star' 22
        Summary 25
    How early? 26
        Summary 29
    False world evidence for real world people 30
        Find the pathway 31
        Strengthen existing development strategies 32
        Seize the opportunity 33
        A bright future 33
        Summary 35
    Becoming 36
        Summary 38
    Forecasting opportunity/the opportunity in forecasting 39
        Summary 44

Finding the 'right patient, right time, right drug segment' 45
    Summary 48
Phase transition 49
    Summary 54

The Value of Deep Positioning 55

The Value Proposition and Positioning 55
    Summary 62
Searching for meaning 63
    Summary 66
The Mechanism of Value 67
    Summary 71
Patients: At the centre, but out of reach 72
    Summary 76
The median is not the message 77
    People, not plots 77
    Summary 80
Share of ear/share of voice 81
    Summary 83
A space in the *real* mind of the customer 84
    Summary 88
People, Not Populations 89
    A need for 'people research' 89
    The value of understanding patient and physician motivation .. 90
    Steps towards understanding patient motivations
    and behaviour 91

- Encouraging behaviour and retaining patients 92
- People research matters 93
- Summary 94

Finding words 95
- Positioning on Mechanism 95
  - Summary 103
- I second (guess) that emotion 104
  - Summary 108
- How you position drugs every day: The Covid case study 109
  - Summary 113
- Pseudo-positioning 114
  - Summary 117
- "Best" 118
  - Summary 122
- Talking a different language 123
  - Summary 125
- What 3 words? 126
  - Summary 129
- The notion of your potion 130
  - The role of categories 130
  - Summary 133
- Time flies like an arrow; fruit flies like a banana 134
  - The role of analogy 134
  - Summary 137

Knowing **138**
    Getting sign off 138
        Summary 142
    Testing 143
        Summary 151

Conclusion **152**

Appendix **153**
    Top 30 Pharmaceutical Positionings 153

Prologue

It is possible that I have done more pharmaceutical positioning than anyone else. I am pretty sure that I have *seen* more pharmaceutical positionings than anyone else and lucky enough to have been involved in many of the most successful positionings of all time. 'Lucky' because the great positionings are typically a combination of the best clients – small, experienced teams – and a great methodology.

We have a motto: "there is no 'I' in IDEA…" That was partly from the early realisation that I might be more experienced than most, but I haven't positioned a single drug. Not solo, anyhow. Instead, the great positionings were built from insights from curious clients, incredible medics, enthusiastic teammates, serendipity and deep craft. But more importantly, the great positionings weren't the ones that ended in a PowerPoint deck, but in everything that the drug then went on to *do*. It is the doing, not the talking, that makes a great positioning.

It took a few great projects, some not so great ones, and a healthy review of the ones we didn't do, to begin to derive an approach that wasn't just a steadfast application of a process someone laid down a while ago. That is one of the biggest risks I see: that we assume there's only one way. I realised we had to be sceptical even of our *own* approach. Maybe we were lucky more often than we thought. At the very least, I should try to understand what tended to lead to being lucky more often. "The harder I work, the luckier I get," as Samuel Goldwyn said.

Plus, sometimes we saw some incredible commercial successes in our industry, and then, when we saw their 'positioning', realised that it couldn't have been the driver. As for the 'positioning' for one of the biggest successes in the last 5 years? Let's just say it failed every test of a good 'positioning'. Whatever was informing their communications clearly had little to do with their market performance. Their very real 'positioning' on market was not the positioning we saw on PowerPoint slides. When we ask ourselves, 'what if we're wrong?', which we do a lot, this is a real counterfactual. Sometimes the Development teams really are right when they think the Marketing 'positioning' is not important.

This journey led to my first book on the subject being revised and Deep Positioning is now a nod to the need to do things not just differently, but better. It is based on the lessons from successful products and unsuccessful ones. It is logical, and energising – it presents a way for the whole matrixed team to work together and look forwards.

It's not a textbook, although there are more words than version 1. It hits the philosophy hard, rather than the process. It is an examination of 'why fish?' rather than trying to teach someone to fish or giving them one to eat.

I am open to challenge, and feedback. Let's treat this like a prototype. Some of the chapters here were once standalone articles, kept in because they fit the overall flow.

Mike Rea,

CEO, IDEA Pharma

Introduction: Why 'Deep Positioning'

No-one ever launched a molecule, and no-one ever will.

What is the difference between your molecule in phase I and the same molecule on the market? It's a trick question: there is no difference. The molecule won't change from the day you take it out of the lab to the day that you put it onto the market. The only difference between those two points is the decisions that you make and the evidence that you choose to collect. The difference between your drug and another company's, when they do hit market? The decisions that you make and the evidence that you choose to collect.

This is the basis for early phase positioning, and it is one that everyone acknowledges – that if you give the same molecule to two different companies when they leave pre-clinical, there is little chance that they'd end up in the same place. The molecule, whichever hands you put it in, will do the same thing forever - biology and chemistry won't change. But no-one launches molecules - they launch products.

The process of developing your molecule as a solution to a problem - finding a position in the market that it can occupy – is positioning. Deep, meaningful positioning. Strategic choices early in the process need to be proactive – back casting from a future market opportunity and regulatory approval should bring questions back to the molecule. The molecule itself does not contain all the answers that a product needs to provide.

Some 50 years ago, a series of articles by Ries and Trout[1] began to promulgate the idea that positioning is 'what happens in the mind of the customer.' According to them, positioning is not what you do to a product, it is what you do to the mind of the 'prospect' (you can already tell so much about their thinking processes, ad men to the core, from that choice of word). That is, you position (place) your product in the mind of the prospect. Since that time, positioning has been the technique in which marketers try to create an image or identity for a product, brand, or company in the perception of the target market. And so it has remained, unquestioned by many in the pharmaceutical industry who failed to understand its consumer-oriented origins. We'll cover this kind of positioning (product or brand positioning) later in the book.

---

1   Al Ries and Jack Trout were marketing strategists and authors who are known for their work on positioning, a concept they introduced in their 1981 book "Positioning: The Battle for Your Mind." Positioning, in their text, referred to the process of creating a unique image or identity for a product, service, or brand in the minds of consumers. Ries and Trout argued that the most successful products and brands are those that are able to differentiate themselves from competitors and occupy a unique position in the market. Ries and Trout also wrote numerous other books on marketing and branding, including "The 22 Immutable Laws of Marketing," "Marketing Warfare," and "The New Positioning." They have been influential in the field of marketing and are known for their practical, actionable advice for businesses looking to improve their marketing strategies. Al Ries died in October 2022.

The idea has its merits. But it is also incredibly limiting. It starts with the view that the product *already* exists, and your job is to find a way to make people want it. After all, if it's just about perception, or image or identity, well we might as well continue what we're doing: let R&D decide on our product, its comparators, its functional positioning and its benefits, and then just develop an ad campaign telling our 'prospects' that it's best in class, to give patients 'freedom to be who they want to be'…

This, unfortunately, is positioning as done by most. So much reverence for the positioning statement, as if the statement itself carried any value (one management consultancy even goes so far as to market 'test' the statement in quant research).

What is missing from that philosophy is clear. It is what comes *before* there *is* a product, (including choices of indication, dose, formulation, endpoints, value proposition, etc.). That is Deep Positioning.

Deep Positioning is, or should be, everything you do. It's a set of decisions that you make, early enough to actually change the *product* you end up launching. Applied correctly, you don't have to worry about which 'mind' you're talking to in your customer group – the regulator, the payor, the patient, the prescriber – because you have actually positioned your product in clear valuable space where they all get the same value. Cymbalta isn't one of the best positioned drugs in pharma because of its statement, but because of a set of very brave decisions made before phase II.

Deep Positioning, done well, is something that every single person developing a product should be able to deliver. It should be tangible in the label, not just hinted at in the detail aid. It is not about what you can say now, but what you want to be able to say in the future.

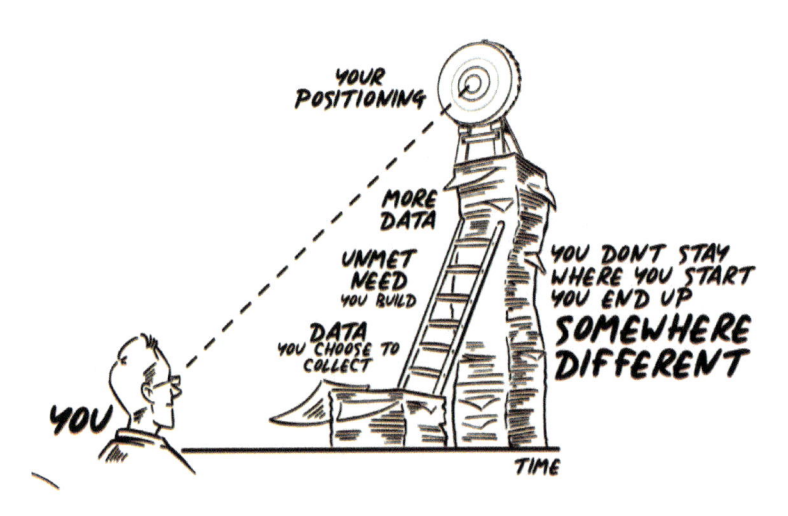

Figure: A diagram showing a person labeled "YOU" looking up at a tower/ladder. Labels on the tower from bottom to top read: "DATA YOU CHOOSE TO COLLECT", "UNMET NEED YOU BUILD", "MORE DATA", "YOUR POSITIONING" (at the top as a target). To the right: "YOU DON'T STAY WHERE YOU START YOU END UP SOMEWHERE DIFFERENT". The horizontal axis is labeled "TIME".

So, why is the Ries and Trout approach to positioning 'dangerous'? Well, the idea that marketers can position any old thing perpetuates a dangerous mindset in pharma. The statistic runs that only one in four launched drugs returns its own investment to its originator, a situation perpetuated by the old thinking. For example, at the peak of the 'statin wars', Pravachol spent a fortune on TV and in print on promotion, all of which drove Lipitor scrips.

You already know you're in a dangerous spot when you're asked to position a drug that's in phase III – apart from the statement and the messages, the room for manoeuvre has gone, the label is all but written. For as long as marketers accept this state of affairs, they'll always be limited to the detail aids and ads with patients staring into sunsets or running down beaches, or held in a perfect balance of efficacy and safety. Products, in this model, will continue to be effectively deep positioned by R&D and then tossed over to marketers to polish the 'perception'. Every marketer who accepts that role diminishes the idea that 'unmet need' is something that marketing should be bringing to the table, that baking competitive differentiation into the product is their responsibility, that the value proposition is a critical part of successful launch.

The Ries and Trout idea keeps marketers as an afterthought at the end of a serious process of 'following the science', and the beginning of a phase of Sales and Marketing. Positioning can be, has to be, something more meaningful than 'the space in the mind of your customer that you want to occupy'. It has to be tangible, has to be built into the product, has to be real and evident in your choices. The product has to be able to justify a place on a formulary, to argue for its value at an HTA, to fight for share in a competitive landscape.

This positioning, Deep Positioning, is what happens *before* the molecule becomes a product. Deep positioning shouldn't be an afterthought. It is a forethought, and the more thoughtful it is, the better.

Summary

- "Deep Positioning" refers to the strategic decisions made during the early stages of drug development that determine the positioning of the product in the market.
- Traditional "product positioning" focuses on creating an image or identity for a product in the perception of the target market, but "Deep Positioning" goes beyond this by making decisions that shape the product itself.
- "Deep Positioning" decisions made early in the development process can change the final product that is launched, and can be seen in the label, not just hinted at in the detail aid.
- These decisions can include choices of indication, dose, formulation, endpoints, and value proposition.
- "Deep Positioning" done well can position a product in a clear and valuable space where all stakeholders (regulator, payor, patient, prescriber) see value in the product.

Who is this book for?

The 'same drug, two different companies' question is not an abstract question. It happens all the time, and with increasing frequency.

Consider these curves. They have been chosen because, unlike many others where there may be a difference between the molecules (statins have head to head data, for example), it would be hard for anyone to argue that any one of the checkpoint inhibitors is 'better' than the other. The difference between these curves is based in decision making – Deep Positioning.

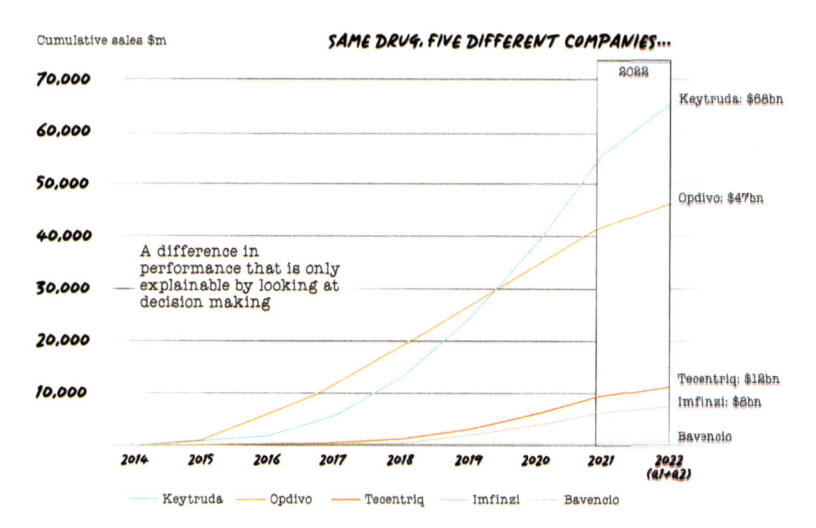

SAME DRUG, FIVE DIFFERENT COMPANIES...

Cumulative sales $m, 2022: Keytruda: $68bn, Opdivo: $47bn, Tecentriq: $12bn, Imfinzi: $8bn, Bavencio

A difference in performance that is only explainable by looking at decision making

The five companies here made choices: which indication, what endpoints, what value proposition. They also made choices about when to launch, or when not to; choices about what to combine with, and what not to. They took decisions when they saw early phase data that showed that these drugs *increased* tumour size, rather than shrank them, based on RECIST criteria.

The argument of Deep Positioning is simple: turning PD1/ PDL1 molecules into great medicines is not something that can be left to 'follow the science'. $70bn difference between Keytruda and Bavencio (both from ostensibly 'oncology' companies) reveals an asymmetry in Development. There may be lots of valid reasons for those decisions of course (portfolio intensity, budget, and more), but still, decisions... Why did Opdivo gain an early lead, and then lose it? Decisions made and not made.

So, who does the Deep Positioning of pharmaceuticals? Development does Deep Positioning, ideally in as multi-disciplinary a team as possible. This will strike many as controversial. The arguments that it is 'too soon' to 'do' positioning often ring out, even in phase II – that, 'if we could just wait until we see the data', *then* we can get on with positioning.'

Development does Deep Positioning, and clearly sometimes does a great, or not so great, job of it. We'll talk later about an ideal team composition in early phase, but as soon as we agree that choosing a market position for a molecule is a discipline based in early phase, we change the dynamics – we appreciate the consequences of choices made and not made, scientifically, clinically and commercially.

So, who is this book for?

It is for teams who want to develop great medicines.

What is a great medicine? First, it is one that gets to patients, and therefore onto market. There is no medicine less valuable than one that falls before that threshold. Second, it is one that outperforms similar molecules – that might be asymmetry in time to market, or a product profile that more closely meets the needs of patients, and therefore creates more commercial value.

Principles of Deep Positioning

1. Deep Positioning *is* about the product
2. Positioning has to do real work, not just be a basis for communication
3. The draft label is the best positioning template
4. Write '…will become' at the top of the positioning, not 'is'
5. Positioning is not about what you can say today, it is about what you want to be able to say in the future
6. Find your 'mechanism of value'
7. Try to crystallise your positioning in 3 words
8. Don't allow people to position your drug who don't realise they're doing so
9. Deep Positioning is not a popularity contest: the best positionings are not always ones that people like
10. The best time to position your drug was in phase I. The next best time is now.

What is Deep Positioning?

Ending up somewhere different

To repeat the trick question, 'What is the difference between your molecule in phase I and the same molecule on the market?' The molecule won't change from the day you put it into healthy volunteers to the day that you put it onto the market, if it gets there.

So, if the molecule stays the same, what changes? Well, you're collecting evidence (learning) and thereby the ability to make claims - from the sterile world of clinical trials, into the messy real world of decision makers. So, it's not the molecule that chooses its market position and viability - it's you.

This is the basis for our philosophy on Deep Positioning, and it is one that everyone acknowledges – that if you give the same molecule to two different companies when they leave pre-clinical, there is little chance that they'd end up in the same place. By extension, these choices are ones that the molecule cannot make – you are back casting from what you think is commercially attractive, and what is feasible to develop.

To address the 'why?' of positioning, it is critical to remember that the word 'positioning' exists as a verb and as a noun. Positioning in phase I (verb) is essential, but it is not critical to have '*a* positioning' (noun). That is, whoever is making those choices about market position and product in phase I *is* positioning the drug even if they don't call it that. Doing what seems obvious at that stage is a decision itself – it will end up passively positioning your drug. To outperform, in development and on the market, decisions in early phase need to be active – the back casting from market and regulator should bring questions back *to* the molecule.

THE 3 LEGGED STOOL OF OPPORTUNITY

Any 2 are not enough!

(Stool legs: APPROVABLE, EXECUTABLE, COMMERCIALLY ATTRACTIVE; seat: TARGET OPPORTUNITY)

This idea, speed to better questions for the molecule, is the basis for the discipline of asymmetric learning.

Knowing that you are always positioning your drug is key. Processes that are intended to produce 'a positioning statement' are unfit for purpose in early phase, because the goal of positioning in phase I and II is to provide options, not a single statement. This is liberating, and it does demand that commercial attractiveness is a key consideration in early phase, not something that can wait until later.

Choosing to end up somewhere different than competition is an option, but one that is lost the longer that 'the obvious' is pursued. 'Follow the science' is a seductive idea, but science itself is enhanced by better questions.

Summary

- The molecule won't change from the day you put it into healthy volunteers to the day that you put it onto the market, if it gets there.
- The ability to make claims changes from the sterile world of clinical trials, into the messy real world of decision makers.
- It is not the molecule that chooses its market position and viability, it is you.
- Positioning in phase I is essential, but it is not critical to have 'a positioning' (noun).
- Choosing to end up somewhere different than competition is an option, but one that is lost the longer that 'the obvious' is pursued.

The 'North Star'

Often in Deep Positioning, we'll talk about 'the North Star'.

The relevance becomes clear when you look up definitions:

The North Star or Pole Star – aka Polaris – is famous for holding nearly still in our sky while the entire northern sky moves around it.

So, what does that mean for positioning? Well, far from being about the way we message at launch, or the form of words we use in the statement, it sets a clear vision as a fixed point, towards which we can travel. Rather than saying how great our drug is in NSCLC, we can focus on its pan-tumour role; rather than its efficacy in gout, we can focus on its role in auto-immune or inflammatory disorders.

If we all know where we're going, the steps and turns we take to get there can be less random, and less focused on the here and now, or the next launch.

This highlights the tension many feel between a 'launch positioning' and a 'strategic positioning/ Deep Positioning' - "we need messages for launch" can feel at odds with a North Star vision for the product.

So, consider two scenarios. In one, you decide to focus on whichever launch someone has decided they are going to deliver, without having had an input to that choice. You make the best of the data that they chose to collect, and you develop a nice wrap for them. Then you do the same for the next launch, and so on, each time just making the best of the data you're given. Out of that might emerge a clear and consistent positioning for your drug if it's easy to see how those indications and messages tie together.

If, however, it was 'a perfect blend of efficacy and safety in Ulcerative Colitis', and 'rejuvenation in rheumatoid arthritis' and so on, you will have created a constellation, a Big Dipper possibly, of messages, ideas and associations. They may shine brightly together, but a) you will have missed the chance to direct the choices of indication, b) the kind of endpoints that would be beneficial for your market position, and c) you will have had to fight each battle on its own terms, without a clear plan for the implementation.

In the other scenario, you decided as a team on your North Star. 'We're going to be the auto-immunity drug' or 'we're going to be the Erythroid Maturation Agent'. With Deep Positionings like those, indication choices become easier, as do choices of endpoint (disease activity, perhaps, instead of pain; anaemia-related disorders, instead of dialysis-related endpoints).

Your messages for each launch might be different, but importantly when you're considering them, they can be consistent and non-competitive - they can build towards a stronger message that cuts through. It is positioning as a pull, not a push. Establishing a North Star will help the market see the constellation you want them to see by strategically connecting the individual stars in the universe of possibilities, rather than gazing upwards into a haphazard and overwhelming cosmos.

Positioning as a pull, not a push

Your data when you launch should be a product of, not the basis for, your positioning.

Just imagine if you'd spotted an opportunity no-one else did to show a benefit on an endpoint that matters to patients, payers and physicians.

Then, imagine that you didn't do that, and instead just embarked on a 'follow the science' signal-seeking path to market.

Imagine those two 'yous'. And decide which is happier about what they did (or didn't).

The North Star allows an active approach to positioning - to build towards where you want to be in 10-15 years, rather than where you are now.

In the end, if you combine the 'pull' that the North Star can provide, you can pre-decide what claims you want to be able to make and pull them through with you. You will have created a draft label for your team, instead of pulling your messages from a label someone else wrote. That approach means you can choose what you want to communicate.

Positioning is about what you want to be able to communicate in the future, not what you can say now

Lots of people confuse messages with positioning. Instead, positioning is about directing the journey your drug takes, and which data it collects, so, when the time is right, you'll be able to tell a blockbuster story with messages no-one else can deliver.

If you chose to just message the data that were collected (possibly just those sufficient for a regulatory approval), you're satisficing - making the best of the situation you are in. The North Star means you can stay consistent, stay focused, while the entire rest of the market moves around you.

Summary

- The concept of "the North Star" is useful in relation to positioning in the pharmaceutical industry.
- The North Star is used as a metaphor for a fixed point, a clear vision, towards which a company can travel in order to create a consistent and non-competitive positioning for their product.
- Having a North Star allows for an active approach to positioning and helps teams pre-decide what claims they want to make and pull them through.
- There is a tension between a "launch positioning" and a "strategic positioning/Deep Positioning" and the importance of considering the long-term vision for the product.
- The North Star can be applied to other industries as well, such as retail, technology, automotive, and food industry.

How early?

At some point, you're going to want someone to want to write a prescription for your drug, someone to want to pay for it, someone to want to take it, and someone to let them.

In your day-to-day, you may see all of that as somewhat distal, something that you can take care of when the need comes to 'sell' it. But each step in that process is something you either considered, or you didn't, back in phase I/II.

Choosing a market position started then. With the choice of indication, the line of therapy, or the patient segment. Perhaps the desired effect size. When the team pursued that one position, they chose *not* to pursue others - it is inevitable. That choice is called positioning - the decision to favour one market position over others. While it may not be the thing you think of when you hear the word, it is positioning. Just consider, in your company, who really chooses market position?

Choosing a value proposition started then. Perhaps the team chose the comparator, the endpoints, the effect size. When they made those choices, they chose a value anchor. It is inevitable. That choice is called positioning - the decision to favour one value proposition over others. While it may not be the thing you think your market access colleagues do, it is positioning. Just consider, in your company, who chooses the value proposition?

Someone decided that they owned 'unmet medical need'. Just consider, in your company, who is in charge of understanding medical unmet need, now and in 5-10 years' time, when your drug might launch?

Someone either wanted to understand, or chose to wait to understand, physician and patient motivations to prescribe and take medicine, now and in 5-10 years' time.

Because of path dependence, choices made early limit the choices available to those tasked down the line: your phase III clinical colleagues may not thank you for the lack of data outside a small population, or any real understanding of dose curve; your regulatory colleagues may not thank you for determining what can go into the label well before they had a say; your market access colleagues may have been grateful not just to vary price based on the label you pre-determined, but to have chosen a value proposition to inform that price negotiation.

Deep Positioning is the discipline of finding a market position. Whatever else it shows up as - the statement, the communication strategy - it is about optimising decision making. It narrows the path to market. But positioning does so actively, rather than passively - when it is making choices, it does so in the full knowledge of its impact on your colleagues.

Active vs passive

The beauty of the active approach is that it harnesses your imagination, your insight, your understanding of why you developed the medicine in the first place. It harnesses your knowledge of where else you could (and will) take the product.

You will never have less evidence than you will at launch (especially for those lifecycle-based medicines – oncology, anti-inflammatory), so you need to be able to provide an idea of where you want to take it.

Lifecycle-based medicines are those where your first launch is just that, and you expect a lot of subsequent launches in new indications or lines of therapy

If your drug is ever to be on the market, this matters. Most processes assume you can do this later. You can't. Your room for manoeuvre was limited, or empowered, by whomever chose your market position for you. And through active, early Deep Positioning, you ultimately expand your ability to withstand changing market conditions and unforeseen circumstances because you have strategically chosen your direction and designed it in a way that provides optionality for all members of your teams to adapt from a unified springboard.

Summary

- The process of positioning, which involves choosing a market position and value proposition, starts in the early stages of drug development and has a significant impact on later stages.
- Choices made during early positioning can limit the choices available to teams involved in later stages of development.
- Deep positioning, which involves actively and strategically choosing a market position and value proposition, can help expand the ability of teams to adapt to changing market conditions.
- Understanding unmet medical needs, physician and patient motivations is important in positioning.
- Most processes assume that positioning can be done later, but it must be done early in order to be effective.

False world evidence for real world people

What is at the core of our biggest challenge in pharma? We develop and present 'false world' data on our drug to regulators, payers and prescribers, who then try to gauge how best to release it into their real world.

The task of Deep Positioning is to find a place in the real world for our medicines, but the route to that point is via a series of 'false world' studies. This chapter is not about the validity of that approach, but to recognise the challenge of *positioning* in this situation.

The 'place in the mind of our customer' goal when positioning pharmaceuticals is a place firmly embedded in the real world: of direct and indirect competitor treatments, of imperfect diagnoses and real, messy patients, of cost, value and incentives. This reinforces, more than anything else, the need to not treat messaging and positioning as in any way interchangeable. In ethical pharma, you can say (only) what you have proved, and if all you have proved is your drug's place in a false, sterile world setting, then the challenge of communicating value in the real world is made harder, by design, and by choice.

Your task, when positioning drugs, is to position for the real world: to provide a direction for Development, to specify data to collect, to decide what you want to be able to claim in the future. The passive approach can only ever reflect a false world view, however nicely packaged the story looks. This active look forward is critical to launching a drug that works when it lands.

Find the pathway

Integrating "real worldness" into the drug development process can be challenging and complex, with a plethora of decisions to make along the way. In the past, pharma has tended to opt for lowest technical risk on the way to regulatory approval which leads to presenting "false world" data to the market at time of launch. This means two things: that real world evidence is often added as an afterthought, and that our messaging can only reflect what we studied. Unfortunately, even the advances made in incorporating comparative effectiveness considerations into the mix leads to challenges. For example, is the standard of care the same in every region?

Finding the ideal place for a new drug, both in reality and in perception, is the key focus for Deep Positioning and this is where real world evidence comes into its own. Presenting an opportunity to understand how best to use a drug in a regimen, in sequence, in severity of disease, line of therapy, etc., and with what concomitant medications.

The industry needs to understand what part of "reality" it wants to evaluate. So whilst it is critical in the pharmaceutical development pathway to obtain "clean" perspectives of what impact a drug has, the industry is essentially leaving patients with fewer options than ideal. For example, it is estimated somewhere around 90 per cent of people living with depression would not qualify for a clinical trial into an anti-depressant, simply because they are too complicated. So, as well as reviewing ways that pharma could look at 'all comers' in development studies, it is critical to also look into techniques to evaluate the data that these techniques would then produce.

Strengthen existing development strategies

Some in the pharma industry have questioned the need to include real world evidence in a regulatory submission. Planning for further evaluation should be a key part of the mix.

In addition, rolling studies over a longer timeline, instead of doing the bare minimum, would be helpful. We have seen companies deliberately stop trials at six months or 12 months because they believe the effect of their drug will wear off after that - that is the kind of thinking that we could all agree we'd like to see die out. Most patients would be keen to know if a drug 'wears off' after 12 months.

With the rise in availability of generic products, among a range of other options patients now have access to, the essential reimbursement hurdle is changing the ways pharma thinks about development. Many companies are starting to realise the need to demonstrate the value of their product against everything else that is available to prescribers. Considerations of real world evidence can kill two birds with one stone.

Another factor to add to this mix is that the difference between the molecule and the product will become increasingly apparent to pharma. The molecule is fixed, but decisions on regimen, device, companion diagnostics, dose, etc., can all contribute to better adherence, tolerability and fitness for the purpose of the medicine. There is a massive opportunity to make drugs more effective simply by altering some of these parameters: If pharma paid as much attention to this side of its business as it does to trying to find new molecules, that could make that same difference to effectiveness. This could potentially add a whole host of innovation to the process.

Seize the opportunity

Pharma companies generally want to create a competitive advantage, whilst prescribers are interested in understanding exactly how and in what ways a new drug will benefit their patients. Instead of companies launching drugs that only beat placebo, and whose effects as part of a treatment strategy are unknown, some companies are looking to be more patient-centric: That means real people, in real world scenarios, with real diseases. Incremental advances are one thing, but companies who preach right patient, right drug, right time have to start to deliver on that message.

A key part of real world evidence is companies actually being interested in what happens to their drugs once they are on the market and being used by patients on an on-going basis. In addition, new devices, diagnostics and other aids to patient management have to start to be considered early in development, rather than as afterthoughts.

A bright future

Imagine, for example, that pharmacovigilance, instead of being a 'necessary evil' was to become an opportunity to constantly monitor drugs in the real world. Whilst pharma might be worried to learn that their drugs are less effective than they want, they may in fact learn the opposite.

There will be key steps taken by some in the industry to simply integrate their activities, so more pooling of data can be done, leading to better statistical assessment of where a drug works, and where it doesn't. Apart from that, greater willingness to explore the possibility of segmentation - finding those patients who don't need the drug, or who don't benefit, rather than trying to pretend it works in everyone. Essentially this means we are likely to see less averaging of populations and more willingness to pre-specify sub-populations.

We can either continue to develop drugs for a false world, and provide evidence of how our drugs work there, or we start to realise the reality of our customers' lives and deliver products that meet those needs.

Summary

- This chapter delves into the challenge of positioning pharmaceuticals in the real world using "false world" data presented during the drug development and regulatory approval process.
- In order to effectively communicate the value of a drug in the real world, the need to integrate real world evidence into the drug development process must be underscored.
- It's crucial to understand what aspect of reality the industry wants to evaluate and how to evaluate the data that real-world studies would produce.
- It is recommended that real-world data should be integrated into the drug development process early on and planning for further evaluation should be a key part of the process.
- In order to obtain more comprehensive data and make better decisions, it is suggested that rolling studies over a longer timeline, instead of the bare minimum, would be beneficial and the industry should investigate techniques to evaluate the data that these techniques would then produce.

Becoming

This one is simple. It builds on a few basic principles:

- You're positioning for the future, not for now
- You're positioning for data that you could collect, not data you already have
- You're using an active framing, not a passive voice

There is so much debate about positioning templates, statement templates and processes, but most of them have a singular flaw: they describe the 'as is' rather than the 'to be'.

Templates themselves are problematic. Statements should be crafted *after* the idea of your drug is decided, whereas they tend to be treated like a recipe card. My drug *is*, so *it does this…* In so many meetings, it doesn't matter how many times you say that you're positioning for the future, you'll keep hearing 'we don't have those data' or 'we can't say that'.

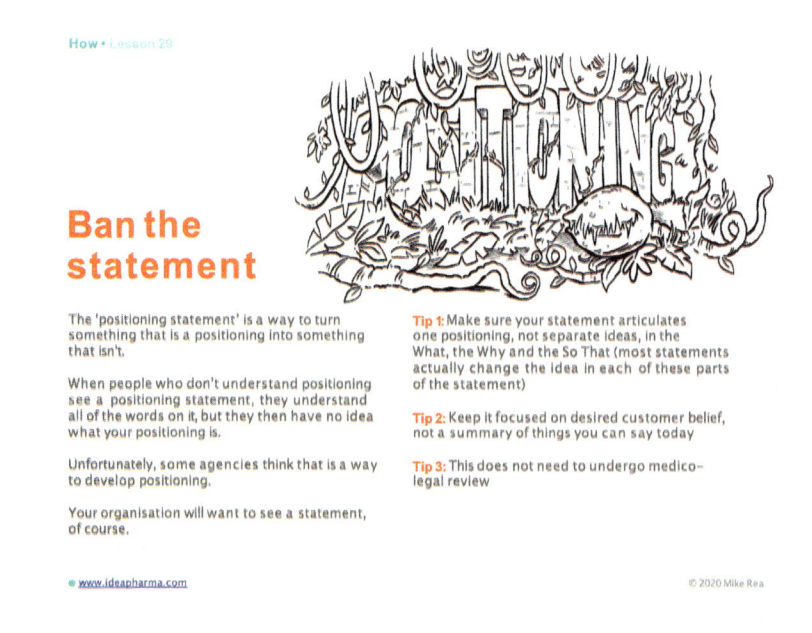

Ban the statement

The 'positioning statement' is a way to turn something that is a positioning into something that isn't.

When people who don't understand positioning see a positioning statement, they understand all of the words on it, but they then have no idea what your positioning is.

Unfortunately, some agencies think that is a way to develop positioning.

Your organisation will want to see a statement, of course.

Tip 1: Make sure your statement articulates one positioning, not separate ideas, in the What, the Why and the So That (most statements actually change the idea in each of these parts of the statement)

Tip 2: Keep it focused on desired customer belief, not a summary of things you can say today

Tip 3: This does not need to undergo medico-legal review

www.ideapharma.com © 2020 Mike Rea

Of all the ways that positioning is done badly (especially by management consultancies), starting with the statement narrows the conversation to passive 'as is' outcomes.

Instead, do this: at the top of any template you choose, write 'Product X will become…'

Once you write 'Product X will become' at the top of your statement, most of what follows can be focused on real positioning - positioning for the future you want for your product, not the data you have from your clinical studies. And for as long as you avoid 'non-statements' like 'best in class' or 'perfect balance', it can provide a clear direction not just for those communicating, but those developing the data.

Deep Positioning is a choice of what you want your product to become, so it's simpler to include that starting point.

If you're positioning for launch, this rule still applies. You may need launch messaging, but limiting your choices to data that you have already collected will lead to two things: limited potency in your messages, and no clear direction forward. Customers will want to know what problem your drug is aiming to solve, not just what hurdles you cleared in Development.

Summary

- You are positioning for the future, not for now
- You are positioning for data that you could collect, not data you already have
- Use an active framing, not a passive voice
- Avoid "as is" outcomes and "non-statements" like 'best in class' or 'perfect balance'
- Start with "Product X will become" when crafting a positioning statement, to provide a clear direction for product development and communication.

Forecasting opportunity/the opportunity in forecasting

Many late stage 'failures' are actually just financial decisions: decisions not to proceed because the business case is not clear. Unfortunately many of those decisions will have been wrong, because the opportunity assessments that supported them were wrong. We know *that* because some of the opportunity assessments that lead to phase III and launch decisions are also wrong, just more visible.

A major opportunity for companies is in opportunity assessment - to deliver competitive advantage by seeing the future differently, and positioning differently. Certainly there's room for seeing the future the same way as others (it's a healthy sense check to understand where they might be stuck), but any creation of opportunity from that vision will still require creativity.

While we all laugh at Decca saying 'no' to The Beatles, or to Steve Ballmer's famous dissing of the iPhone, they were not wrong at the time - there was a lot of work left to do by the band or by Apple. But the work that was left to do was *part* of the opportunity.

In pharma, there *is* no accurate forecast in pre-launch - they're all wrong, and wrong in both directions equally. Even McKinsey will admit that. Yet, those forecasts are relied upon to make decisions. So, the question is: if they're not accurate, are they useful?

FAMOUS FAILED PREDICTIONS
* Decca/Beatles: "We don't like their sound, and guitar music is on the way out."
* Steve Ballmer/iPhone: "There's no chance the iPhone is going to get any significant market share."
* Thomas Watson (IBM)/home computers: "I think there is a world market for maybe five computers."
COULD YOU FORECAST THE OPPORTUNITY FOR THIS MOLECULE ACROSS ALL CANCER INDICATIONS?
WE CAN'T SEE ANY OPPORTUNITY FOR THIS ASSET...
MIKE REA + Rbolusso20

False negative forecasts of opportunity will kill early-stage assets. Unfortunately they tend not to ever be counted as such - who would ever know? They're shelved along with the asset.

False positive forecasts will kill late stage assets by letting them launch and fail on market. So, the number isn't the point. The point is to understand the work that is to be done... If the commercial evaluation was to reveal *how* to make something successful, and to shape the product, it becomes a useful input, not just an enabling number. This is a reasonable description of pharmaceutical positioning, when it is done right.

Deep Positioning means looking in places that others are not looking, or looking at the same places differently - to take 'assumptions' as just that, and to unpack what is assumed and what is known. Applying the same methodologies as others do to the same evaluation will lead to the same wrong answers. The sources of error in the forecasts are well known: asking the wrong people the right questions, asking the right people the wrong questions; being unable to successfully translate a TPP to an actual product description, or assuming that, even if you did, the people you're talking to can imagine themselves five years in the future making decisions in the presence of alternatives and competitors, and price decisions and more.

While there might be comfort in being wrong with the crowd, rather than right and isolated, that is not what companies should be incentivising. Applying imagination to commercial evaluation is key. How would your company feel, today, if your analysts were the only people who could see opportunity in a blue ocean? How much validation would you require? Critical to the Viagra case study wasn't seeing the side effect, it was in learning quickly that it could be something valuable (in a space that had no precedent).

The key to forecasting opportunity is not in knowing that there's a trend away from 'guitar bands', but in providing a series of 'ifs': if they can write their own songs, if those songs are good enough, if they can work hard and do promo, if they stay together... This is also true of pharma. But the 'ifs' should come from the opportunity assessors, not be handed down from the product developers who'd prefer to keep things simple.

As we have seen so often, the 'medical unmet need' that kicks off a Development exercise might turn out to be already met when the product gets there. There are dozens of drugs that look very similar, because they followed the same traditional path, in a disease like Multiple Myeloma.

In another example, even when a cure for Hepatitis C was launched in 2013/14, calculations at the time suggested that 260,000 treated people per year would essentially rid the US of Hepatitis C. The market saw a peak of only 160,000 people initiating treatment in 2015, with a decline ever since... Estimates are that there is more Hepatitis C in the US today than ever. The gap between goal and shot is getting bigger year by year. A cure might seem like the best possible connection of solution and unmet need, you might find that non-obvious competition exists (under-diagnosis, lack of interest in treatment, too high a price, etc.).

Key here is to bake in the positioning 'if' to the exploratory phase, to inform Development plans. Some of Development has to be to find out, not just to press the 'ahead' button.

If, instead of saying 'yes' or 'no' to The Beatles, those companies had had the opportunity to engage in exploratory options and questions, before they made a decision, things might have been different. In pharma, that option exists (the band is already optioned), but only if those who understand the market are able to *describe and design* the opportunity, not just estimate the opportunity pre-described by the Clinical team.

The opportunity in forecasting opportunity is to involve forecasters in design, not just prediction, to understand the variables and the invariables, and to be part of the conversation about the product in development. The part of the opportunity that sees the work that can be done is a fundamental asymmetry for companies to embrace.

Summary

- There is a major opportunity for companies to gain a competitive advantage by seeing the future differently and positioning themselves differently.
- Many forecasts in the pre-launch phase of pharmaceuticals are inaccurate, but they are still relied upon to make decisions. The question is whether these forecasts are useful or not.
- False negative forecasts of opportunity can kill early-stage assets, while false positive forecasts can kill late-stage assets by allowing them to launch and fail on the market.
- The key to forecasting opportunity is not in knowing that there's a trend, but in providing a series of "ifs" that come from opportunity assessors, not product developers.
- Applying imagination to commercial evaluation is key. Companies should be incentivized to look for opportunities in places that others are not looking, or to look at the same places differently.

Finding the 'right patient, right time, right drug segment'

The famous idea of a monkey trap is captured nicely by Wiktionary

1. *(literally) A cage containing a banana with a hole large enough for a monkey's hand to fit in, but not large enough for a monkey's fist (clutching a banana) to come out; anecdotally used to catch monkeys that lack the intellect to let go of the banana and run away.*
2. *(figuratively) A clever trap of any sort, that owes its success to the ineptitude or gullibility of the victim.*

That rather neatly captures the problem, when positioning, of the 'niche' challenge... Typically someone somewhere in the organisation will be afraid of 'nicheing' the drug - giving it an addressable market that is less than it 'deserves'. This is often driven by over-ambition (or over-forecasting), rather than a good assessment of the positioning and the realistic opportunity for the drug. Like the monkey, the marketer may see the whole market as desirable, but in trying to gain hold of it is either stuck or has to drop it, gaining nothing.

MONKEY TRAP

MARKETER TRAP

HMM, I DON'T WANT TO "NICHE" MY PRODUCT.

MIKE REA + RGalvão 22

Pharma seems an unusual market in seeing a 'niche' as a bad thing. The presence of a foothold, or an ideally-suited part of the market may seem like opportunity is being left to others. However, it is hard enough to find a place that really does need what you have, without trying to spread a positioning so wide that it appeals to everyone - better to stand for something than to stand for nothing... If you realistically gain 10% of a market, your company will thank you for it - much better to take aim at that 10% from the beginning, rather than gaining 5% of the whole market after a whole lot of heartache.

The anti-niche marketers are also often fans of passive, instead of active positioning (or of the idea that data will do the work). However, in terms of your strategy, most successful products start with a niche and expand - it isn't automatically limiting. Rest assured, if someone offers you a niche when you're climbing a rock face, you'd take it happily. The same should apply to your positioning.

Summary

- The monkey trap is a metaphor for a situation where one is tempted to capture a large market but ends up trapped due to your own ambition.
- In the pharmaceutical industry, some marketers view a niche market as a drawback.
- Niche markets are better suited to the company's offerings than trying to appeal to a larger market.
- Successful products usually start with a niche market and expand later.
- A niche market should be embraced as it provides a foothold for the company.

Phase transition

Deep Positioning means displacing something... It might be a direct competitor, indirect competition, or established medical practice, but something somewhere has to be edged aside for your product to fit into a future landscape.

Two kinds of phase transition problem impact innovation in pharma. The first, perhaps one of the biggest single killers of innovation: the emphasis on phase transition rates in Development. Groups organised to simply get from phase I to phase II or from phase II to III have a perverse incentive structure. To create a truly agile or innovative organisation would mean breaking that paradigm.

The second is a failure-to-plan that impacts launch.

As an analogy, while we can all imagine that the world would be better off with a single design of power socket, the phase between then and now is almost unimaginable (for fans of the irreverent and status quo-challenging animated show *South Park*, this is known as the 'underpants gnomes' problem: 'collect underpants - something - profit', where 'something' is undefined). It is uncrossable territory, as there is too much to change. It almost doesn't matter how much you plan your two states, your 'as is' and 'to be' - the magical world it could be if everything changed overnight. Shifting the user base is not going to happen because the gain (to them) is less than the hassle. Imagine being forced to change all the sockets in your house, or your business, because someone decided the UK plug was the standard for safety, reliability and more.

Pharma misses this challenge when it launches cell and gene therapies, or new classes more broadly - it often presents the 'to be' data as compelling enough to change practice, but forgets the practicalities of shifting the user base. There are many ways in which clinical studies are unlike the real world, but the motivations of investigators rarely track to the complexities of real physicians and real patients (and their very real payer systems).

If I want my gene therapy, for example, to succeed in a world in which you, the prescriber, today have a perfectly viable enzyme replacement therapy, I need to do more than just provide an equivalent outcome over four years at an equivalent cost, on a standard measure. This has been true of PCSK9s vs statins, novel heart failure medicines over standards of care, and more. It's particularly true of cell and gene therapies, which on the surface seem magical, but encounter real-world logistics and economic problems.

The user experience of the installed base matters - and the user experience is more than simply 'prescribe and watch'. The day 0 experience (deciding, prescribing, getting it paid for), the day 1 experience (watching for some form of outcome, seeing how the patient is doing), the day 30 experience (is it different in any way?), etc., all matter. How much does the prescriber get paid to do what they did yesterday vs what they'll get paid now? How much does their comfort with established practice factor in when there's a totally novel approach to learn? What does the patient experience that they wouldn't have experienced with the previous gold standard?

In the cartoon, this traditional linear 'think about the user later' approach is lampooned - everyone recognises this pattern, but if patients, physicians and payers are thought about too late, the risk of delivering something they don't want, or that doesn't fit, is high.

Your technology may well be better, on paper. But prescriptions aren't written on paper - they're written in the minds of the prescriber and the better you understand their experience, the more likely you are to launch a successful alternative. Too much time on the data, and not enough on the practicalities, can kill great medicines.

Unfortunately *this* problem can find its source back in the first problem of phase transition. Science can win at each phase transition until approval, when real user experience takes over. A programme that wasn't designed against the end user experience (value endpoints, duration, ease of application, economics and more) will then have an army of sales, medical affairs and more trying to force fit it into a real world application.

Many companies that think of themselves as 'learning organisations' miss this problem, preferring their learning to be linear. 'If we take care of the science, the rest will take care of itself' is a way to avoid positioning in early phase, but just creates problems for later.

Data that are easy to collect may be at odds with, or insufficient for, evidence that will effect a phase transition in the market. Ironically, data that are easy to collect continue to drive the phase transition rate problem in more established pharma companies - for as long as it's easy, and you are rewarded for simply moving from phase II to phase III, it's hard to shift to a more productive system. Your positioning may require endpoints that need to be validated in phase II in order that they can be part of your phase III, and therefore your label, or patient identification/ stratification that needed to be part of a pre-planned approach. Wishing that you'd had those endpoints at any point after the start of phase II is too late, but it happens every day.

The evidence that may make a physician move away from familiar territory to unfamiliar practice is unlikely to sit simply in the standard chart for oncology studies, the Kaplan-Meier curves.

Kaplan-Meier curves are a graphical representation of survival data over time. They are often used to visualise the probability of survival for a group of individuals or patients over a specified period of time.

In a Kaplan-Meier curve, the x-axis represents time and the y-axis represents the probability of survival. The curve is constructed by plotting the number of individuals still alive at each time point and then connecting the points with a line. The curve is typically used to compare the survival probabilities of different groups, such as those receiving different treatments or those with different characteristics such as age or gender. Unfortunately, although they are a great way of showing data from a clinical study, they offer little other than probabilities for an individual patient at the start of their treatment journey, or for their physician.

Recognising the real world in which decisions are made is the starting point. Even for an index patient, all the pros and cons of a new kind of decision need to be understood and planned for. So many great drugs fall down for their failure to predict established behaviour's response to a request for change. If you'd have reasons not to change your power sockets (network voltage, service providers, difficulty in finding parts, appliances that would not fit, etc.), all of those reasons and more can be found in a new drug or modality. So, your positioning can't be 'this would be better, but we recognise you have to change your practice...'. Anticipating what needs to change, and how to help people transition, is critical.

Summary

- Deep positioning refers to displacing something, whether it be a direct competitor, indirect competition, or established medical practice, in order for a new product to fit into the future landscape.
- Two kinds of phase transition problem impact innovation in pharma: emphasis on phase transition rates in development, which can lead to a lack of agility or innovation in organizations, and failure to plan for launch, which can impact the adoption of new technologies by users.
- When launching cell and gene therapies or new classes more broadly, pharma often presents the 'to be' data as compelling enough to change practice, but forgets the practicalities of shifting the user base.
- The user experience of the installed base matters, and the user experience is more than simply 'prescribe and watch'. It includes factors such as the day 0 experience, day 1 experience, day 30 experience, etc.
- A technology may be better on paper, but prescriptions aren't written on paper. Too much time on the data and not enough on the practicalities can kill great medicines.

The Value of Deep Positioning

The Value Proposition and Positioning

No student of English could fail to see the similarity between the phrases 'benefit statement' and 'value proposition'.

Here, it is argued that the value proposition is the more useful concept and should be attracting the energy and attention of organisations in early phase to a far greater degree.

While there is disagreement, even among 'marketers', about when positioning should be done, and what it means, no sensible voice can be raised against the value proposition being a fundamental design choice to be made *ahead of* planning phase II and certainly III. For this reason alone, perhaps, there is an argument for replacing 'positioning' with value proposition, especially in phase I/ II.

But consider the additional truth: most 'positionings' in most companies omit any statement of value, instead attempting to link a feature/ benefit with a message (and even worse, a message that the audience in research 'liked'). However, it is harder for a value proposition to exclude a positioning, even if by default.

This chapter argues that, in the presence of a strong value proposition, positioning is secondary. The value proposition connects the academic world of discovery with real users of medicines. Drugs are moved through pipelines because someone somewhere thinks they might bring value to the lives of patients or healthcare providers - rarely, however, is that value articulated early in development, with a consequence: the drug's purpose is subsumed, so it becomes 'for rheumatoid arthritis' rather than what it was 'for' *within* rheumatoid arthritis - its purpose is weakened or assumed. Fortunately, there is a hard measure of whether the value proposition is there - the medicine is reimbursed and included in guidelines. Without one, or with a weak one, it is not. Unfortunately, the value proposition is still often regarded as something that can happen close to launch - perhaps because of confusion with 'price'.

(It would be argued further that value should be decoupled from pricing or 'market access'. While linked, they are not the same: value is as linked to clinical development and marketing as it is to pricing. Price can be set relatively close to launch - value proposition cannot.)

Deep Positioning offers the opportunity to design drugs patients want and need. Outside the clinical markers we have traditionally measured, there are obviously many other things that affect somebody's health and health behaviours. What's important to a patient may be different from what is important to a pharma company, payer or provider. We need to think about what matters to patients so that when a product comes to market after 10-15 years, patients can make an informed decision with their HCP based on outcomes most important to them.

Patients have become more aware of the value of their data to help develop better products and services. How do we equitably engage them in the drug development and innovation process? Having a relevant value proposition early/earlier in development may only become more important in a future where patients will own their data and select to share it with providers and companies who add value to them.

Let's agree that the choice of the word 'statement' or 'proposition' is an arbitrary one - they're essentially interchangeable, in English and in use here. So, let's unpack 'value' and 'benefit', as they are the key words.

Value: the worth of something, the importance of something, the usefulness of something, the point of something... These are all calculations of benefit. It is important to consider that, in the way it is used in a 'positioning', the 'benefit' is (simply) the highest ownable benefit. Similarly, the 'value proposition' is supposed to be a statement of the highest-known value of a medicine to an audience. Value depends on quantifying a qualitative benefit (where 'benefit' is defined as the connection between a feature and a goal of the audience). So, it is evident that a proposition of value and a statement of benefit are essentially the same thing, with one working at a higher order. It articulates the answer to the question: what's the point of your drug (to a customer)?

Let us also acknowledge that 'positioning' was a statement of benefit that made more sense in the day when choosers were not following the guidance of institutional buyers or guideline makers. It is no longer acceptable to say that a drug is 'the drug of choice for physicians who want to do the best for their patients' (apologies to all the companies who have written exactly that for half of their positioning statements). Now, a medicine has to prove its value, its worth, in the clinic. If 10% improvement on a scale versus placebo is of value to the patient because it allows them to do something they couldn't otherwise, that is different than simply stating that 10% as a self-evident benefit.

Which of these outcomes is more compelling from an Alzheimer's Disease study - 50% increase in ADAS-Cog over standard of care, two point increase in ADAS-Cog, or avoidance of nursing home for one year? They may all be the same (if Aricept is believed to deliver a four-point average effect, providing an additional two points on a very wide scale is allowed to be called '50% increase'). One outcome has clear value - the others need to be extrapolated to a 'valuable' outcome. Having a statistically significant effect on 'negative symptoms' in schizophrenia is easier than proving the clinical value of doing so. But one of those choices will pay far greater dividends on the market.

We also need to consider that the worth of something is not self-evident, even to users, on first exposure, and rarely, if ever, from presentation of a set of features on paper. There is a strong rationale for more appropriately guiding 'market' research in this area. More work should be done by the developing company to prosecute an argument of value, rather than waiting to hear how a (necessarily) poorly-informed audience member might characterise it. If you believe that your new PCSK9 lipid lowerer will bring additional value over statins, and over the other PCSK9s, then it is for you to construct that argument and show how you might show it... Clearly there are different kinds of value it will bring to different patients, but then it remains for you to prove so. If your new gout drug will make a difference to progression of a degenerative disease, it remains for you to show that it does so, not simply to do what is needed to pass regulatory scrutiny. If you were asked 'what do you think is the point of the iPhone 13?' and showed you a set of features, or said 'what do you think of a device that allows you to do x?', which do you think is more likely to prompt a useful response? Fortunately, knowing what your drug might do, or how it might be of value, is something that you should be uniquely equipped to do.

Note that the phrase 'value for money' introduces a cost-effectiveness measure into a judgement of value. While value may be fixed in a given scenario, the 'for money' part of the equation is variable.

Consider: how much do you use Google, or Wikipedia? Most people use Google several times per day and consider it 'invaluable' to allow them to do their jobs. When asked how much they would pay to use Google, however, the normal answer is 'I wouldn't...' Even though most respondents know that they *do* pay for Google indirectly, any suggestion of direct payment would now strike them as 'too much to pay' for that value. Similarly with Wikipedia. (That pharmaceuticals could benefit from such an arrangement seems far-fetched now, but it did when Google was launched, too.)

At a conference on market access once, the audience were asked how many had flown business class to get there. With many hands raised (the pharma companies, one presumes...), They were then asked the question: 'could you describe the value to you of flying business class?' Of course, many kinds of value were given: space to work, better ability to sleep... When asked the follow-up question: "And, if you were paying for the trip yourself, would you have made the same choice...?" - the number of hands fell substantially... So, with the same *value* to those business class seats, the 'for money' part of the equation changed the outcome when it was the subject's own money at that same price. (This simplifies the complicated value model that the airlines understand: the status, the lounge access, etc., but the truth holds.)

PJ O'Rourke once neatly pointed out the rather sharp observation that it matters whether you are spending your own money, or someone else's, and it then matters whether you're spending it on you or someone else (it simplifies massively: that 'someone else' being your 3 year old daughter is not the same as if it is a drug-abusing repeat offending criminal in another county, of course). While self-evident in retrospect, even this simple model suggests the reason that some brands can push price if they appeal to individual rather than institutional buyers.

So, it is clear that it is value *for money* that should determine price. It is also clear that the 'value' part of the equation should come first. Consider the following kinds of value proposition for a drug that impacts cognition:

- Additional cognition to AChEis;
- The Short-Term Memory drug;
- The Orientation drug;
- The Dementia drug (pan-indication);
- Nursing Home Avoider.

Despite deriving from the same molecule (with different development plans), no payer would regard these as of equal value (and therefore equal value for money at a given price). These are propositions, not just calculations, of value. There is a value *story* to each of them.

These choices are readily identifiable, and directly connect the goals of audiences with features of a product. Understanding the value a market might place on a product is the role of design: the case study of the Toyota Prius and the Honda Insight is interesting here. An assumption that people who wanted a hybrid were doing so to save money led to Honda developing and marketing their Insight as cheaply as possible, whereas Toyota launched the more upmarket Prius at a higher price with a view that people who wanted to show that they cared about the planet wanted to be seen to care, and didn't want to give away too much luxury in doing so. Not only did Toyota charge significantly *more* per car, it also made more *margin* per car, and *outsold* the Insight by many orders of magnitude.

When pushed, most companies admit that their product will have variable value - that its 'value' delta over the current standard of care is greatest in high risk patients, for example, or in patients with higher performance status. Averaging out the value delta across all populations can only decrease the value you take to payers.

What could be the arguments against value proposition displacing positioning? 'Positioning is the place in the minds of your audience that you want to occupy' was the old mantra. The value proposition handles that challenge with ease, but makes it a more fundamental part of the product you develop. Being able to articulate what the point of the product you're developing is becomes as unarguable as making sure it is safe. If you occupy that place in reality, through Deep Positioning, it is a lot easier to occupy that place in your audience's mind.

Summary

- The value proposition is a fundamental component of positioning and in this scenario, a more useful concept to work with, and should be given more attention in early phase of product development.
- Positioning is often done incorrectly, and mostly omits any statement of value.
- A strong value proposition includes positioning, and is a fundamental design choice to be made ahead of planning phase II.
- Value should be decoupled from pricing or 'market access'.
- Value depends on quantifying a qualitative benefit and answering the question: what's the point of your drug (to a customer)?

Searching for meaning

What do we mean when we say 'meaningful'?

There may be disagreement about the role of the 0.05 p value in 'statistical significance', but it's easy to ignore that the 'significant' part of that phrase does not mean 'meaningful', but rather means 'probably true'.

There are also bookshelves devoted to the scientific definitions of 'clinically meaningful'. Papers will discuss the thresholds of minimum detectable difference (MDD) and minimum clinically important difference (MCID), and attempt to derive a 'clinically meaningful' threshold. Let's be careful with that term here - it does not mean that it is a difference that matters to a patient, just that it represents a noticeable, and apparent, difference in a measure such as Global Rating of Change Scale (GRS), where a patient feels something as vague as 'improvement'.

Tests like the *six minute walk test* (6MWT, how far a patient can walk at sub-maximal effort in 6 minutes) have the value of being relatively easy to perform, although with wide variation in daily results for most patients. You or I might manage 500m in six minutes, but a patient with heart failure might do 300. It isn't a heart failure-specific score, of course, as COPD, MS and a lot of other conditions would reduce exercise capacity.

So, it is a thing we can measure. When we're Deep Positioning, we need to consider more: does it matter, and does it differentiate?

For many patients with heart failure, a walk of 300m is a considerable undertaking. Two flights of stairs might be inconceivable. Their adaptation may well mean they cope by not doing either of those things - increasing the downward spiral. So, if a drug increased that 300m to 330m, most physicians would say that was a meaningful change in the patient's status. But let's put that onto the stairs. It might be 11 steps instead of 10. We could argue about the utility of that extra step, and it's entirely possible that an agency could craft all kinds of nice ads about that last step, but we then need to decide whether the 10% is the point. Could it be about the first step, not the last? The willingness to start, rather than the ability not to stop so soon? The anxiety, the sense of improvement vs decline, the walk to the car instead of sitting in front of the TV? So, to the question, 'does it matter?', the answer might well be 'it depends.'

Then, to the thornier issue: does it differentiate? Because it is easy to measure, it is easy to mandate as a regulatory endpoint. It isn't essential (even for heart failure), but most regulatory groups internally will tell you that 'the FDA says you have to…' So, you collect the data, against placebo, and then you find that you're playing the game of numbers - is my 12% better than your 13%, given that we had different NYHA status patients in our studies? How much, then, is that extra 2 metres worth to a payer, or a physician, or a patient, over a generic, or over standard of care?

If you know this is a problematic conversation for your Commercial colleagues to have, when should you have begun to think differently? It is entirely possible that the endpoint that does measure the value your drug brings uniquely could have been used in your phase III (alongside a 6MWT). But, it should then have been validated in your phase II. Which would mean that you'd considered it before designing that phase II programme. A drug that 'works' in heart failure is not the ask, even if it works well enough to be approved. A drug that changes people's lives meaningfully is the task, and to fulfil that task, we have to search for a shared version of 'meaningful.'

Summary

- It is critical to understand the concept of "meaningful" in the context of medical research and treatment.
- The term "significant" as used in statistics does not necessarily mean "meaningful" in terms of real-world impact.
- It is important to understand the "clinically meaningful" threshold, which is often used to determine the effectiveness of a treatment; it does not necessarily mean that it matters to the patient.
- The *six minute walk test* is a great illustration: while it is easy to measure, it is not an endpoint which will poorly differentiate between generic and novel medicines.
- The goal of medical research should be to find a drug that changes people's lives meaningfully, and that this requires a search for a shared understanding of what "meaningful" means.

The Mechanism of Value

Here's the thing about 'Efficacy'. It doesn't exist. It is an abstract concept, a placeholder, a shorthand, a broad clustering of 'things that a product does', a descriptor under whose vagueness all kinds of good things can hide.

There is only one time in a Market Research presentation when 'efficacy' should be capitalised, and that is on the slide that shows "Efficacy: what doctors meant when they said they want more of it…"

Unfortunately, that is wishful thinking. An overwhelming majority of primary research still comes back reporting that doctors want more Efficacy, Safety and Tolerability, and ideally at lower cost. Usually that research will have been framed by what your audience thought they could have more of.

Imagine telling a salesman that you want a car. "What kind of car?" "One with more, well.., efficacy. One that does what cars do, but better…" Or, imagine presuming that people buy new mp3 players because they want to hear their mp3s better.

If we take something as simple as obesity, all of the following parameters could be included in a review of Efficacy:

- degree of average weight loss at a certain timepoint (the usual understanding),
- effect on satiety,
- effect on 'food addiction',
- durability of response,
- effect beyond withdrawal,
- amplification of lifestyle change,
- effect on visceral fat,
- effect on mood or mental appetite,

- effect on fat distribution (visceral vs subcutaneous),
- responder rate…

Each respondent in a piece of Market Research study on obesity may have meant one or more of those dimensions when they chose 'Efficacy'. It isn't enough to think that we're talking about the same thing – why not go one step further, and find out *exactly*?

This is all unfortunate, for it is in the detail that efficacy becomes beautiful, and strategic opportunity becomes possible. Consider the simple switch that enabled Lipitor to gain advantage: moving the market from measurement of outcomes to measurement of a lab test as a primary measure of efficacy. Lowering LDL is a measure of efficacy, as is lowering major cardiac events - physicians were ready to believe that one of those led to the other, however long the chain between the two.

The problem comes when we all assume we're talking about the same thing when we use the word (unfortunately, unlike some other languages, English doesn't allow a way to hear whether we mentally capitalised the 'e' or not…). The Development folks hear that physicians want more Efficacy and think, 'well, that's fine… Let's go looking for something, anything, in the studies that fits under that banner…" That was the case for the statin market, where ever-escalating outcomes studies proved they reduced serious events. But, luck and judgement led to Lipitor seeing things differently, saving money, time and risk.

Worse, an acceptance of the use of Efficacy ends up with a view that your product, from all the products out there, offers the perfect 'balance' of Efficacy and Tolerability (a net clinical benefit that just happens to favour your drug). Consider an anticancer drug for a moment. Efficacy is often taken to mean Overall Survival, or Progression-Free Survival. Those are perfectly reasonable things to measure. However, there are many other ways to evaluate the efficacy of an anticancer agent: visible effect on tumour regression, effect on symptoms, effect in different lines, in different stages, in different risk patients. All of those dimensions are running through the minds of oncologists, who see patients in their real worlds. So, for a marketer to claim their drug offers 'a perfect balance' will often come across as more than a little thoughtless.

WHAT DOES A 'CANCER DRUG' DO?

©2021 IDEA Pharma

ORR: Partial response, total response (RECIST)	Reduction in total tumor burden	Reduction in burden of other cancer therapies	Reduction in need for other interventions (e.g. reduction of raised intracranial pressure)	
Complete response, stringent CR: 'undetectable disease'		PFS2 (time to second progression)	Reduction in burden of other therapies (e.g. analgesics)	
Lower use of supportive or palliative care	Time to progression			
		Progression-free survival	Delay in need for next line of treatment	Overall survival
Disease burden	Time to recurrence/ treatment failure	Minimal residual disease	Disease-free survival	MRD (minimal residual disease)	Cure (25 years disease-free survival)
		Relapse-free survival			
	Delaying progression of disease state		Metastasis-free survival		
		Magnitude of circulating tumour DNA (ctDNA) reduction		Persistency of ctDNA reduction	
	Specific biomarker measures (blood and body fluid-based)		Time to metastasis (Metastasis free survival)	Event-free survival	
Pain reduction and prevention	Sustained employment	Improved and sustained HRQoL	More quality time with loved ones	
	Improving and sustaining functioning		Reduction in psychologic sequelae (depression, anxiety)	
Sustained reduction of symptoms (e.g. breathlessness)	Less hospital visits	Sustained benefits on ecological momentary assessments	Palliative/end of life efficacy focused on comfort, QoL, and pain management	
	PROs (fatigue, appetite, anxiety, etc.)			

PROXIMAL   **DISTAL**

The problem of that capital E is manifest in any review of 'Unmet Need'. (The same rule applies to those capital letters...) Because we can all agree that Efficacy is a shorthand for a granular set of (often conflicting) things that a drug might do, any review of unmet need must respect that granularity or be rendered pointless, a waste of Energy, Enthusiasm and Effort...

Even worse is that clinical trial endpoints can be assumed to be the best measure of 'efficacy', ignoring the adage: not everything that matters can be measured - not everything that can be measured matters. So, our provable, claimable 'efficacy' can often be reduced to whatever statistical significance was shown in studies. The thing that matters might not have been measured, but we now have to make what was measured matter.

This is at the core of the challenge of Deep Positioning. If done early enough, the wonderful connection between something your drug does and something your audience want (even if they don't know it yet) can be included in studies, and claims can be made. If it is done late, your task becomes to create a belief with flimsy evidence and real endpoints that don't measure value.

Summary

- Efficacy is an abstract concept that is often used as a placeholder for a variety of positive attributes of a product or treatment.
- In market research, doctors often express a desire for more efficacy, safety, and tolerability, at a lower cost.
- The problem is that the term efficacy is often used too broadly and without a clear definition of what it means.
- To truly understand what physicians mean when they ask for more efficacy, it's important to drill down into the specific dimensions of efficacy that they are thinking about.
- The use of the term efficacy can lead to a false sense of balance in a product's net clinical benefit and can be thoughtless in the context of anticancer drugs.

Patients: At the centre, but out of reach

There's a relatively old joke in pharma, that it's hard to have the patient at the centre of everything you do when the quarterly revenue target is already sitting there. But that hasn't stopped 'the patient' being held out as the new North Star, the *sine qua non*, the singular purpose of what drives pharma these days. Wander the lobbies of the top 30 (and, one suspects, probably the next 300), and see what look like the same images of 'patients' and their 'stories' staring back at you (presumably created by the same agencies/ PR companies each time...).

Do some people believe it too? A couple of minutes in a meeting a few years ago still resonate. When asked which, *exact*, 'epilepsy patient' the company had in mind now that it had put the word 'patient' in every single sentence they uttered (given the tens of epilepsies and the thousands of ways it can present), the room responded with 'you know, THE epilepsy patient'...

It is well known that averages hide all kinds of nonsense. But no nonsense so dramatic as to conflate the experience of a rare disease baby with a permanent seizure condition with someone who might have had two seizures in their lifetime, of unknown origin... And to average out all of the others into '*the* epilepsy patient' and their generic 'journey'. The next question was asked by one of the best people in the room: '*how many people here have met people with epilepsy?*' About 2 or 3 hands went up (from around 35 people in the meeting). '*But we are supposed to... The company has encouraged us to...*' As it transpired, most of the people in the room had reasons they hadn't: compliance, training limitations, etc. They weren't wrong - they had certainly been inhibited. But, as that experience showed, it wasn't the law stopping them - it was interest and intent.

And that remains the paradox. The word 'patient' is heard constantly (with a silent recoil from many an audience). But it is also known that the real worlds of patients, of humans with disease, are unexplored. A remarkable lung cancer patient shared her observations: *"pharma is certainly interested in my tumour, and making sure I bring it to the trial site when I'm supposed to, so they can measure it... But interested in me? Not so much..."* And she told such an important truth then - that her world remains unexplored even to those who have her in their sights (and in their sites...). As a human, rather than someone, a subject, a *patient*, with a tumour. She went on to add that pharma was interested in her advocacy, but not, of course, in helping with her rather harder questions about treatment effects, or aspirations, etc. We do like our clinical trials to be sterile, to ask small, closed questions. But the real world is not sterile - in that world, people have emotions, logistics, goals that may not include tumour shrinkage, cost constraints, mess... They have other drugs they need, or don't want; they have people to care for, and to care for them, with all the understandings and misunderstandings that brings.

And then we come to our biggest problem. So afraid are we of ourselves that we have tied our own hands behind our backs, and we start with the assumption that we're there to sell people drugs. So, if we want to talk to a patient (heaven forfend), we must approach them as an 'advocacy group', so we're not perceived as trying to market to them. Never mind if we really want to hear or understand their experience - the system is set up to feed clinical trial recruitment, to push ever more eligible patients (that word again) into our trial pipeline, as if getting into a trial is some kind of unalloyed good.

Has there ever been a cogent defence of DTC? It is vapid, boring, and, given that it says nothing (and does that badly), hard to imagine that it is even driving adoption these days. But we have created a paradigm where legal and compliance and others are very interested in what we say to patients.

But, instead of *talking* to patients, what if we started listening? Not just to what they 'need', but what else they want? There are so many incredible groups out there now - potent patient groups (MMRF, MDA, Savvy Patients and more), who want to talk to us, who want to share. But, start to listen, and there will be a hundred hands that go up to stop you doing that.

You can't talk to patients. You *can't* talk to patients. You can't *talk* to patients. You can't talk to *patients*. You can't talk *to* patients. It doesn't matter how it's said, there are a hundred reasons you can't.

What you need to remember is that there are a million reasons you *should*. In their world of want, there are so many drug and non-drug solutions that could drive your company. And you can't assume someone at your company has ever heard them, or that those patients, those humans, have ever been able to ask for something that would help. The fact remains that there are ways to do what you want - to listen, to understand, to hear - and that you may need to re-assert the reasons to do so. Although the river has been running one way - to prevent you getting near a patient - in recent years, you can fight your way back upstream, and it is only right that you do so. Those rules, held up to stop you doing so, are often invalid - held up because it is easy to say 'no', but hard to say 'yes'. You can, with purpose, do better.

By the time you've defined someone as a patient, you've already broken the frame. To reduce their humanity to their 'patient-ness' and then to further reduce it to the frame of how they could fit your pre-conceived label or segment, may be convenient, but it dilutes your purpose if you really believe that they should be at the centre of what you do. Their richness, their goals, could be part of your framing. Consider this life stage graphic - it's easy to start to believe that this one person is 'a patient' but even this timeline suggests that there is so much more happening at each stage... It's up to you to find out. And the people who want to stop you doing so are either not hearing you ask loudly enough, or have misunderstood your 'why'. Both of those can be fixed.

Summary

- The pharmaceutical industry claims to put patients at the center of everything they do, but often relies on generic images and stories that fail to capture the unique experiences of individuals with different diseases.
- Many people in the industry have limited contact with actual patients and the industry's focus is often on clinical trial recruitment rather than understanding patients' needs and experiences.
- The industry's use of Direct-to-Consumer (DTC) advertising can be vapid and boring, and not effective in driving adoption.
- The system is set up to feed clinical trial recruitment and push patients into the trial pipeline, as if getting into a trial is an unalloyed good, but this is not always the best thing for the patient.
- The pharmaceutical industry should focus on understanding patients' experiences rather than trying to sell them drugs and comply with legal and regulatory requirements.

The median is not the message

People, not plots

It is easy for pharma to see its data the way that the regulators and statisticians have to. To prove that a drug worked is critical. But, it does not mean that should be a default way for you to communicate.

Individual physicians see individual patients, and they make decisions. They're not choosing for 500 patients or 5,000. It is even possible that they see only one or two of the patients you're most interested in, in any given year.

Consider how dehumanising the Kaplan-Meier plot can be, as discussed earlier. At the top of these curves, in an oncologist's practice, a decision will be made, discussed, negotiated. Whichever treatment choice they make, the patient with breast cancer might die in a week. Whichever treatment choice they make, the patient with breast cancer might not die in five years. Yet too many pharma companies would shout 'but look at the data! It is clear which product has the best efficacy!'

DECISION **RIGHT?** **WRONG?**

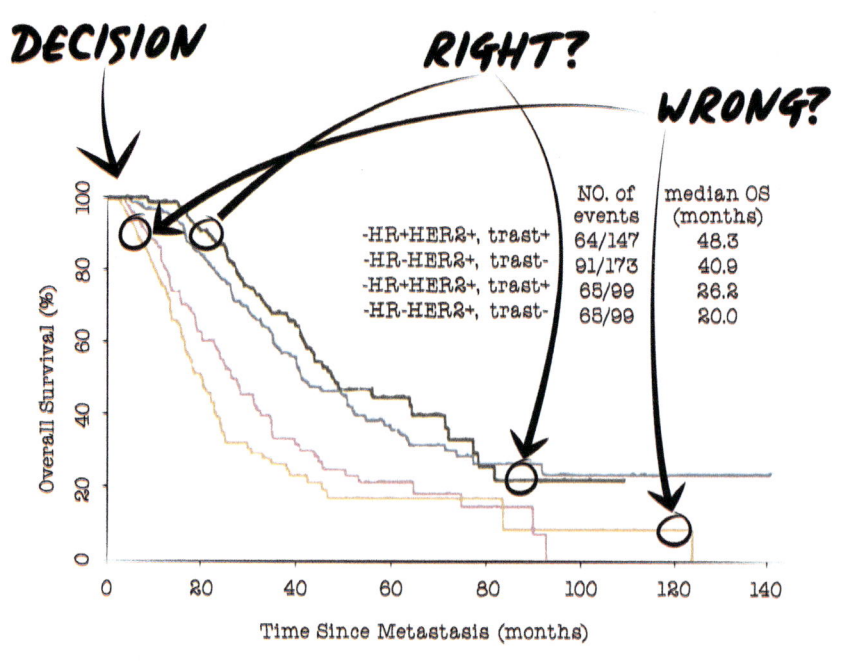

	NO. of events	median OS (months)
-HR+HER2+, trast+	64/147	48.3
-HR-HER2+, trast-	91/173	40.9
-HR+HER2+, trast+	65/99	26.2
-HR-HER2+, trast-	65/99	20.0

Those circles were real people, with real lives. Certainly one could look at a plot like this and hope that a patient might have 'the right kind' of cancer, and we might know that the chances for a woman taking trastuzumab are better in the early months and years. But for any one patient, you might never know. Even with a plot like this, you might never know, for that one person.

Pharma tends to talk to physicians as if this is literally the only thing they or their patients should care about. 'Ignore the costs, ignore the visits to the physician, ignore the side effects. Look at the efficacy! Look at that median!' All other things being equal, we'd expect them to go with the treatment with the best evidence, especially if it has head to head evidence. However, it is arrogant for pharma to believe that it knows what 'all things being equal' might mean to *that* physician, and *that* patient.

Often, of course, there is no head to head. Often, there will be no guide to managing side effects (are they transient, or treatment-limiting?). Often, there will be issues not just of cost, but hassle in getting a drug approved by payers.

You can't start your positioning just by trying to benefit ladder eight months of median OS. No physician in practice will ever see 'overall survival' - it is a study metric. No physician has a 'median' patient. You trivialise the differences treatment might make to a patient if you assume it is only about time, whatever the cost in impact on daily life and financially.

You might think a decision at the top of a curve like this is obvious. But real people do make different decisions. If you think they're wrong to do so, you might need to check your understanding of their lives.

Summary

- The median is not the message: It is important for pharmaceutical companies to prove that their drugs work, but this should not be the default way of communicating with physicians and patients.
- People, not plots: Individual physicians see individual patients and make decisions based on their specific needs and circumstances. They are not choosing for a group of patients.
- Dehumanizing: Kaplan-Meier plots can be dehumanizing as they reduce patients to data points and do not take into account their individual lives and experiences.
- Arrogance: It is arrogant for pharmaceutical companies to assume that they know what is best for the physician and patient and to ignore factors such as cost, side effects, and access to the drug.
- Real people: Real people make different decisions and it is important to understand their lives and circumstances before making assumptions about their choices.

Share of ear/share of voice

If your voice isn't getting through, it is possible that making it louder is not the solution.

It is typical to hear companies talk about share of voice, but when your message is disconnected from the real goals of your customer, it really won't matter how loud you are, or how much you crowd out other voices - it won't be more relevant with increased transmission.

We crafted a '6 As' model of communication a while back, describing the stages that a message might take from the time it leaves your mouth to the time something good happens.

- Attention
- Awareness
- Acceptance
- Agreement
- Action
- Advocacy

At each stage there, relevance is key.

So, knowing that a message needs to be relevant, surely the predominant idea is to listen, and to hear the customer. Imagine that your customer tells you something she has told no other company, or she tells you a day earlier than she tells your competitor.

So many good drugs land and fail to connect. It is important to consider that it might not be the drug, but the message. Your customer may not be able to give you a better message, but they can give you a better insight or problem statement for you to work with. They might give you a better understanding of the challenge of moving from what they do currently to what you want them to do.

It isn't captured as a metric, but I'd bet the farm on 'share of ear' being a better predictor of market performance than 'share of voice'.

Summary

- Making your voice louder may not be the solution if your message isn't getting through.
- Share of voice alone doesn't matter if the message is disconnected from the customer's goals.
- The '6 As' model of communication (Attention, Awareness, Acceptance, Agreement, Action, Advocacy) emphasizes the importance of relevance at each stage.
- Listening and understanding the customer's needs is crucial in crafting a relevant message.
- Share of ear may be a better predictor of market performance than share of voice.

A space in the *real* mind of the customer

It's a safe bet to say you have things you think that you don't tell people and suspect you have things you believe that you have never thought about, behaviours that are unconscious, or sub-conscious. 'Don't believe everything you think' is a wise statement...

Prescribing is a negotiation, between patient and prescriber. It might be that neither gets what they want, but a lot of the power in the relationship sits with the prescriber - until the patient leaves the room, of course. And, like any negotiation, it is unlikely that each will share their preferences, their goals, or their views of the other.

Classic exercises, such as having marketers draw speech bubbles around the heads of patient and physician reveal that we all know that there is a lot that isn't shared, and a difference in how the illness is presented to the physician, and how it is perceived. No-one asking the question, or giving the answer, believes the 'I think it's about 21 units per week, doctor' response.

The oft-repeated trope that positioning is about the space in the mind of the customer, therefore, runs into a reasonable question: is it their real mind and thoughts, or the ones they want to say out loud? If we accept that we'd like to get to their real mind, we also have to accept that our way to uncover those thoughts needs to be more mindful.

It would be nice to believe that our drugs enter a world of perfect information, and positive motivations. In that world, we could focus our positioning on positive, motivational, aspirational ideas, without needing to educate anyone on mechanism, or disease manifestations. But you don't believe that, and it isn't true.

Motivation can take many forms for a prescriber, but it has to start with understanding their real world. As I wrote earlier, the work to meet your customer where they are and where they want to be has to be the starting point. Your SWOT strengths may be of little relevance in this world, so we have to work harder to understand what is the most compelling part of your proposition, your competitive advantage over what else is available. Importantly, that may not be obvious from your label, your TPP or your future dataset. Some of what your customer does is 'away from' or 'avoidance' behaviour, or is a convenient shortcut to a decision. Not all of their day job is uplifting, aspirational and emotionally positive, even if it looks that way in traditional market research.

As I covered in the chapter ***I second (guess) that emotion***, emotions are elicited, not stated or asserted.

To the annoyance of many, no pharmaceutical product has ever had an 'emotional' positioning[2]. They may have had, indeed should have had, some emotional component to their communications. But positioning is different. Positioning needs some differentiation if it isn't to be pointless. When you pick a place in the mind of your customer, you may also pick getting to what your customer actually thinks about their patients is critical. Many techniques are much better at that than the traditional research: anthropology, psychology and more might reveal real behaviour, and real beliefs, instead of the ones you'd expect. Often we'll find that physicians have the exact opposite real view than the one they give when they're allowed time to craft their answers. For example, it is a truism that some physicians prefer IV to oral for financial reasons, or that they know their patients won't comply, but still see it as a discharged responsibility—but it is a hard thing to have anyone admit in research.

Getting to what your product actually could do for that real customer is not then a trivial exercise. You have to start with that

---

[2] *A lot of brands have an emotional positioning statement, of course. That is not the same.*

real understanding of their real mind. And into *that* mind you can test various ideas of your drug. Guess what happens next? Even the most motivational idea in the world can then come up against negative reactions - from the pharma company marketing team (especially if there is a confusion between messages and positioning), and from the way they are tested. It is hard to like some of the best positionings ever, but liking a positioning is not a good test of a positioning. The point of positioning is not to find nice things to say, but to frame the product in a way that is both memorable, but motivational (as well as the other checklist items: sustainable, unique, etc.). The real source of competitive advantage needs to be dug out and tested against this real customer.

Great positioning works with truths. Often they are very simple truths. The challenge is that they may be inconvenient truths when aired, but it is our job to air them and to work with them, or we risk just saying nice things to people who don't really care.

Summary

- There exists a divide between what customers say they think and their true thoughts and behaviours.
- The power dynamic in prescribing can make it hard for patients to share their preferences and goals with the prescriber.
- Question the concept of "positioning" being left to marketing, as it can be challenging to truly understand the customer's thoughts and motivations when the product already exists.
- Alternative methods, such as anthropology and psychology, can help gain a deeper understanding of the customer's true thoughts and behaviours, as traditional market research methods may not be effective.
- Even with a thorough understanding of the customer's thoughts, a product's positioning may still encounter resistance from marketing teams and other stakeholders.

People, not Populations

People think they are making rational decisions regarding their health and the treatment of any health conditions, but they may be wrong. Indeed, it is common for people to wrongfully predict their own behaviour. This doesn't mean to say that people are naturally hypocritical. Instead, it simply reflects the fact that many of our behaviours are driven by unconscious stimuli or motivators. Therefore, in order to understand their patients better, the pharma industry need to concentrate more on the individual rather than the "market." To this end, behavioural psychology can be invaluable.

A need for 'people research'

Our markets are collections of individual people. People are made up of emotions, unconscious drivers of attitude and behaviour, and a conscious, rational outside. If we can understand this at an individual level, we can understand markets, and develop ways to move markets. It is, therefore, critical for businesses to have an understanding of the unconscious stimuli or motivators that can truly drive the behavior of each individual in their marketplace.

Some of the early ideas associated with behavioural psychology were exclusively focused on the understanding and modification of the behaviours of those with psychological problems. However, predicting behaviour in general and identifying behavioural stimulus is highly useful and applicable to any study that involves people, and that includes consumer markets and patient groups within the pharma industry.

People research is more valuable than market research. Although population studies and clinical trials provide large amounts of data about the 'what' of behaviour, they rarely answer the 'why' of the behaviour, and consequently fail to include the actual motivators that can predict a desired behaviour (i.e. medication adherence).

It is easy for companies to forget that there is a massive difference between evidence at a population level, and a decision for an individual patient made prospectively. The fact that physicians are human, and not simple algorithmic robots, is at odds with the view taken by many pharma companies that just showing enough data will be enough to persuade.

Research has shown that, often, physicians don't communicate what the drug is, what it is for, what its side-effects might be, how best to take it, and for how long. In most cases, they communicate two of those five. Patients sometimes only get one of these five, or a fraction of it. With limited information cascaded down to the patient, pharma companies rely on clinical trial results to convince patients to change their behaviour and adhere to their product. However, this is of little use.

The value of understanding patient and physician motivation

Within pharma, behavioural psychology can help gain a better understanding of what is important to patients. Market research can attempt to provide answers, but its results can be of little value if they can't provide insights that translate to actual revenue. Instead, the real behaviour of consumers, in actively responding to an advertising campaigns' call-to-action and making a purchase, is what constitutes tangible income for a business. Pharma companies can apply behavioural psychology concepts in order to better understand the motivations that influence patients' decisions regarding which pharmaceutical company to trust, purchase from, and be loyal to.

There are a number of factors that influence the personal opinions of both patients and physicians regarding medication and treatment plans. The roots of behaviour, and therefore decision-making, have a much greater influence than most physicians or payers acknowledge. For example, look at how few rationally derived treatment guidelines are actually and absolutely followed by physicians. Factors such as experience, expertise, trust, and familiarity must be playing a part in decision-making at the individual level, and patient by patient.

Internal and unconscious factors can influence an individual's opinions. Indeed, it has been observed physicians will make different decisions even between two ostensibly similar patients, based on parameters such as perceived optimism, or in one study we did, perceptions of 'laziness' by the patient. Market researchers don't always include such factors, like a physician's personal judgment about a patient, in their population studies.

Additionally, although market research can identify the individually stated goals of patients and physicians, results don't always reveal the true motivators that have led to those goals. It has been observed that in many or most cases, the goals of physician and patient are the most important predictor of the eventual treatment plan, and yet those goals are often interpolated from some deeply subconscious cues.

Furthermore, although treatment planning is a shared activity between patient and physician, their goals might not be the same, and unconscious factors can continue to influence the achievement of either person's goals. Treatment goals are seldom shared between the physician and patient. A physician is balancing their own goals and the patient's, often without knowing that they are doing so.

Steps towards understanding patient motivations and behaviour

Understanding the psychology of decision-making is critical. Pharma companies need to have an understanding of the concepts of behavioural psychology to aid in identifying and analysing such factors, and consequently, predict patient behaviour.

There are two basic steps towards understanding patient motivations that predict behaviour. The first step involves acknowledging that pharma companies are dealing with collections of individual people. Recognise that at every level – regulator, payer, physician, and patient, for example, not to mention every other individual involved such as nurses and carers – they are dealing with people.

Step two is about seeking the answer to the question, "Do they know why they do what they do, and are they prepared to tell us?" The answer to this is critical because people tend to make decisions that aren't solely rational. For example, it can't be tenable that an FDA panel can vote 11-4 on a drug, and that every one of those 15 was only looking at the evidence. Each of the 15 made an individual decision, based on their own feelings about the risk-benefit balance, on the confidence they have in the data, etc. The context within which people decide might even be invisible to the individuals themselves. Biases, prejudice and preferences are all swimming around the central question – subconscious rather than conscious and rational.

Encouraging behaviour and retaining patients

Behavioural psychology can also be applied by pharma companies to entice and retain patients. It is a tremendous opportunity to move from simply looking at what people need to looking at what people want.

Getting to the core of what drives patients to act can have a dramatic impact on healthcare systems as well. Understanding motivations to seek treatment earlier, to arrive at a concordance with the physician on best treatment choice, and to optimize chances of success on treatment, could, for example, make more difference than a new drug that offers incremental benefit. The potential cost-saving impact of influencing factors that motivate and increase the desire to prevent disease among patients are clear.

People research matters

In an industry such as pharma, it is important to recognise that although the market is made up of people who supposedly make rational decisions, these people are actually unconsciously motivated by factors such as emotions, temperament, preferences, and personal opinions. It is equally important for pharma to identify and understand such motivating factors.

Population studies and clinical trials can reveal only half of the story about patients. In order to provide precision medicine, behavioural psychology must fill in the missing half. Patient journeys, pain points, what ifs, micro-environments between patient and physician, and opportunities to persuade and tell stories - these are all places where people research matters more than market research, where opportunities come from understanding, not speculation.

Overall, an understanding of behavioural psychology can aid pharma in identifying deep-seated motivations that may be invisible to both physicians and patients. Companies can then influence the treatment-related decision-making of physicians and the health-related decision-making of patients.

Summary

- People may not make rational decisions regarding their health and health treatment, and can wrongfully predict their own behavior, which is driven by unconscious stimuli or motivators.
- Understanding individual's unconscious stimuli or motivators is critical for the pharma industry to understand their patients better and move markets.
- Behavioral psychology can help gain a better understanding of what is important to patients and physicians.
- Internal and unconscious factors can influence an individual's opinions and decision-making.
- Pharma companies need to understand the concepts of behavioral psychology to identify and analyse factors that predict patient behavior, by acknowledging they are dealing with collections of individual people and seeking answers to why they do what they do.

Finding words

Positioning on Mechanism

There is a myth: 'you can't position on mechanism'.

It is wrong. It is a claim that is clearly falsifiable when you examine the positioning of many of the most successful drugs of all time. For every Enbrel ('rejuvenation'), there is an Avastin (mechanism) or a Losec (mechanism) or a Prozac (mechanism).

It has been heard a lot, and for a long time, yet where is the source of the belief? There are suspicions:

- Ad agencies have convinced many that 'positioning' has to be 'emotional'. This is a clear misunderstanding of positioning - there is a strong argument that ad *campaigns* have to find some emotional resonance with their audience (although evidence for that is also hard to find), but positioning sits upstream of ad campaigns (and medical education, PR, etc.). The myth that positioning has to be 'emotional' will be addressed later. Given how hard it would be to make a mechanistic positioning 'emotional', it is easy to see why they might say it is to be avoided. As we'll see later, this ignores the role that the prescriber plays in the positioning - being spoon-fed emotions may undervalue the drug, and will certainly undermine your chance to tell the story.
- Market Research enjoys their forced-choice ranking approach. "Doctor, do you think 'mechanism of action' is more important than 'saves lives'? No? Oh, so that means you don't think we should position on mechanism…" You

can't market research your way to a positioning, and this is one of the reasons - that's not the way that positioning works. You can't simply take a list of features and choose the most 'important' - that turns the choices into goals, and it would be an odd physician who wanted to say out loud that mechanism was more important to them than outcomes.
- Marketing Excellence: Unfortunately myths tend to repeat as religious beliefs, and most marketers in pharma have learned their process from ad agencies or market researchers. So, the problems from the first two bullets become reinforced in practice.

What do we mean by 'mechanism' anyway? The word itself just means 'part of a process', so it is critical we see it as such - a process with a start, a middle and an end. Imagine saying 'we can't position on process' to someone who drives a Tesla, eats organic and prefers streaming TV to cable.

At a basic level, there is 'mechanism of disease', 'mechanism of effect' and 'mechanism of action'. Let's unpack:

- Mechanism of disease
  - This tends to become a little 'chicken and egg'. What is the 'mechanism' of cancer, for example? You may have hypotheses, which inform(ed) your Discovery, but it would be an odd scientist who suggested any disease had only one pathway. Were you to suggest that one of the mechanisms of cancer is that it commandeers blood vessels, that would be true, if not the whole truth (it would ignore escape, metastasis and many other hallmarks of cancer). It is rational for your scientists then to target that mechanism of the disease. In saying this is a mechanism of cancer, you are not saying it is the only one. Were you to say that it was *the* mechanism

of cancer, one might reasonably expect that a drug that stops the process would also stop cancer.
- When it comes to positioning, Mechanism of Disease is a part of your story: here's how the disease works, here's where our drug stops that happening, and here are the good things that result.
- Mechanism of Disease therefore supports a strategic therapeutic intervention. 'Rheumatoid Arthritis is an inflammatory disease. Drug X is anti-inflammatory' is not the same as 'Rheumatoid Arthritis is an auto-immune disease. Drug X targets auto-immunity.' Both may be true, but your choice of which story to tell is what positioning is for. (RA is also a 'pain disease', a 'mobility disease', etc… So your choice of the idea of your disease is critical.)
- You can see, therefore, why Mechanism of Disease is both essential (how could you tell a story without it?) and important to your choice of positioning (how could your role in disease otherwise be presented?).
- In some diseases, the mechanism of disease is a lot more tenuous, or speculative. If you take an illness like depression, there are hypotheses such as 'chemical imbalance', or the 'amyloid hypothesis' for Alzheimer's Disease, and because of the complexity of the actual disease, physicians may well go along with your hypothesis and internalise it as 'how' the disease proceeds. In fact, establishing the 'serotonin story' was fundamental to the success of 'SSRIs/ SNRIs', even if it wasn't (and isn't) true/ the whole truth.

- Mechanism of effect
  - With the importance of the Mechanism of Disease, the Mechanism of Effect becomes crucial. What do I mean by Mechanism of Effect? Consider how you frame the

therapeutic intervention your drug offers. The effect of your drug on the disease is central to your positioning - it either *is* your positioning, or it explains your positioning. It is either the outcome that people believe your drug offers, or the reason to believe it.

- For example, how does aspirin 'work'? Let's consider: as a 'pain killer', you might consider it an 'anti-inflammatory' (mechanism of effect) and therefore its COX-1 inhibition (mechanism of action) would follow. Should you be thinking of its cardiovascular effect, you might consider it an 'anti-platelet' (mechanism of effect). The *effect* you want to offer determines your *mechanism* of effect - it would be odd to use its 'anti-platelet' descriptor to explain its effect on a headache.
- Your effect can be proximal or distal. For example, you could explain *how* statins work in prevention of MACE, but it is easier to explain how your statin impacts LDL (its proximal effect), and let your 'lipid lowerer' provide the benefit via the reduction in LDL. This may seem obvious, but in the heat of battle, many statin marketers lost their way and started either positioning on emotion (long-term outcomes), on mechanisms that didn't play out ('pleiotropic effects' was a belief at BMS about how pravastatin did 'more than just LDL', which led to one of the craziest head-to-heads ever seen - the PROVE-IT study). Cardiovascular disease is clearly complex, but simple stories ('lower is better', 'how low can you go?', 'numbers of patients to 'target' in 30 days') won the day. Lipitor, a fifth to market 'me too' dominated the (at the time) most lucrative market ever with a mechanistic story, derived from a feature (potency and wide therapeutic window). It was an entirely rational play.
- It may be that you understand the disease a lot better than your customers. Their understanding of the disease

does proceed via new news, new metaphors on how the disease they treat works. Is eczema inflammatory, allergic or something else? With the emergence of new drugs, physicians update their internal stories. Physicians need explanations, to buy in. They also need explanations for their patients ('this is why I chose it, this is why you should take it') to set expectations. Positioning can be explanatory - it frames why the results you see in the data happened, even if the data superficially look similar to someone else's.

Mechanism of effect vs mechanism of action. Describe your outcome, not your input

Most people misunderstand 'mechanism'. There is 'mechanism of disease' (the way the disease works) and 'mechanism of effect' (how my drug affects that disease). Going into the weeds of molecules and targets is rarely a good thing. The 'mechanism of action' should be chosen after you've done your positioning - which 'action' exactly are you describing?

If you say 'Avastin is a VEGF-inhibitor' and I say 'Avastin is anti-angiogenic', you've used a mechanism of action and I've used a mechanism of effect. The second is a lot more useful to your customers.

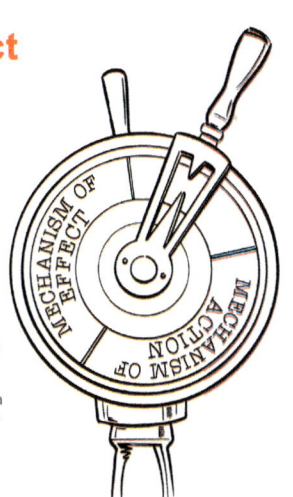

- Mechanism of action
    - How does paracetamol/ acetaminophen work? How does metformin work? We know *that* they work, but very little about how. (The same might be even more true of something like Vitamin D.) Instead we rely on 'anti-pyretic' or 'anti-diabetic' as mechanisms of effect or simple declarative statements. With hundreds of pathways culminating in fever, pain or diabetes, the precise mechanism of action of these drugs might never be known, and may be irrelevant. Drugs can

have 'unknown' mechanisms on their label (modafinil, for example) - empirical evidence will trump an absent biological target.
- Some drugs have different *actions* at different dosages. It is hard, therefore, to claim a single action from any drug. Instead, the 'mechanism of action' should be derived from the 'action' you want to communicate. This might sound counter-intuitive to many: 'surely it *is* a TKI/ CD73/ angiotensin-2 antagonist? That's what the R&D team are calling it.' Well, yes, it could be. Each of these is true, but not the whole truth, for any molecule. Being 'a TKI' is not your mechanism of action - all your drug does may be inhibiting tyrosine kinase everywhere it comes across one, but that action is different in CML and breast cancer. Is its inhibition all you want to talk about, or is it the antigrowth effect, the effect on stability, etc? How do you bring in components of EGF, SRC or more? Not only do you have to choose an action or combination of actions at some point (so why choose the default?), but you're looking for a role in a process. You certainly shouldn't be bound by lexicon someone chose before you got there - they may have had different strategic considerations, or none at all.

There is clearly a lot more that sits in these three considerations - the main takeaway is that 'mechanism' is a variable - it is not fixed by the science, and should be chosen as a part of your storyline.

Why and when would you choose to position on mechanism, then?

Let's consider Avastin, easily one of the most successful oncology drugs ever. Positioned on mechanism for most of its lifecycle (and certainly throughout pre-launch from first indication), in one of the most 'emotional' diseases, this is a simple counter-factual to anyone who claims 'you can't...' As 'a VEGF inhibitor', Avastin spent a lot of its life not being developed. When it was eventually developed, it provided a breakthrough efficacy in colorectal cancer. That was clearly a story worth telling. But that story comes along for free with the approval. But, other things mattered:

- Provide an explanation for its anti-cancer effect
  - Was it by 'chemo-sensitising', 'chemo-sparing', increased permeability of blood vessels, 'normalising' vasculature, or some other 'mechanism of effect'
- Explain why not to expect an effect as monotherapy
- Explain why Avastin would and could have a 'pan tumour' effect in lifecycle, so not change positioning tumour by tumour...

...among other tasks. So, instead of stopping at 'using a first in class VEGF-inhibitor leads to new hope for patients' a simple story of 'cancer generates new blood vessels, a process called *angiogenesis (mechanism of disease, new lexicon for the market shaping)*; Avastin is *anti-angiogenic (mechanism of effect)*; used early and in combination, Avastin offers the hope of longer survival' was chosen (I paraphrase, obviously).

A rational, mechanistic positioning allowed oncologists to bring their own benefits to a new tool. Such was the strength of the IDEA that 'turn off' angiogenesis ('turn down angiogenesis' was the allowable external message, despite the 'switch' visual being seen everywhere), led to blood vessels being the central and only visual theme for many years, through many launches. Even today, *'the anti-angiogenic'* would be given to only one drug, despite there being a lot of 'anti-angiogenics' out there.

Whether you *should* position on mechanism is something we'll cover later. The key thing about positioning is to consider a range of approaches, based on a thorough understanding of the discipline, and choose based on fitness for purpose. Living within the imaginary guardrails of 'you can't' is a limiting belief.

Summary

- The belief that you can't position on mechanism is a myth and is clearly falsifiable.
- Ad agencies may have convinced many that positioning has to be emotional, which is a misunderstanding of positioning.
- Market research's forced-choice ranking approach doesn't work for positioning.
- The word "mechanism" means "part of a process" and is critical to understanding positioning.
- There are 3 types of mechanism: mechanism of disease, mechanism of effect, and mechanism of action. Each plays a role in positioning and tells a different part of the story.

I second (guess) that emotion

As mentioned previously, and in earlier writings, no pharmaceutical product has ever had an 'emotional' positioning[3], to the annoyance of many. They may have had, indeed *should* have had, some emotional component to their communications. But positioning is different. Positioning needs some differentiation if it isn't to be pointless. When you pick a place in the mind of your customer, you may also pick a place in their heart. But if you start with the heart, and forget the mind, you don't have a positioning, you have a vapid tagline

There are *lots* of emotions. (There are even lists of them, dozens of constructs and hundreds of emotions). The problem is that there are fewer emotions than there are products. In choosing an 'emotional' positioning, the less experienced / misled marketer is, by design, choosing a generic positioning. There are even fewer positive emotions, and in restricting the choice to *positive* emotions, the marketer is choosing from a pretty limited pool – unsurprisingly one that has been fished thousands of times before.

Agencies and these marketers are made for each other. Agencies (especially the brand name agencies) would never dream of 'positioning' on anything other than 'optimism', 'trust' or 'joy'… Genericism by design… It is hugely frustrating to see positioning statement after positioning statement take the unique features of a great product and turn it into 'makes the doctor feel like a hero/ guardian/ partner.' Of course, if you have Coke's marketing budget and its reach, feel free to take a tilt at making doctors feel like heroes when they use your brand.

---

[3] *A lot of brands have an emotional positioning statement, of course. That is not the same.*

Something well understood outside of pharma is positioning based on negative emotions: anxiety, desire for peer respect, fear of doing wrong. And in fact, many of the best pharma brands have had this as part of their effective positioning. Lipitor's positioning as the most effective statin understood that while doctors talked outcomes and saving lives, they acted on proximal markers – LDL coming down reliably and rapidly was much more of a concern in day-to-day practice, and that was how Lipitor was positioned (essentially 'the most effective LDL lowerer'). Products that promise to stop your patients coming back have traditionally done better than those that promised partnership, unless there is a financial incentive to keep them coming back (the drivers of behaviour often seem contradictory to the professed goals of the physician – witness the way that payments for IV infusions change practice country to country).

Emotions are great places to go looking for insights. But they are not a landscape against which to position a drug – they are a paint-by-numbers picture rather than a Monet. Brands do have emotional connections with their audiences, but they are/ should be positioned rationally.

The opposite of rational is not 'emotional', it is irrational.

When people say that positioning 'must be' emotional, it is a clear sign that they are conflating both consumer and pharmaceutical positioning, and communication/ messaging with positioning. A rational positioning can easily inform an emotional message ('if I can lower your LDL faster in 30 days on this one drug than any other, a) I'll look like a better doctor, and b) you might live longer and see your grandchildren graduate…'), whereas an emotional positioning ('for those who want to see their grandchildren graduate') has to tie back to something rational in the product. Even if you were able to drive an emotional connection between grandchildren and your drug in the minds of patients via DTC, they'd then come up against the prescriber choice, the payer choice and a whole lot more.

That's hard enough if your drug is the one that *has* those outcomes, but imagine if it isn't. Imagine if you're trying to position on emotion while someone else has gone uber-rational. You don't have to imagine - those case studies exist everywhere (Pravachol vs Lipitor, for example, where a lot of the Pravachol marketing in 1998 drove Lipitor scrips). People will always work their way mentally down the 'benefit ladder' to how your drug might possibly do the things you claim, so you'd better be ready with that supporting evidence.

Too often, a desire for an emotional positioning corrupts the process. It often ends in 'stigma', 'confidence' or 'hope', with nothing at the core. Features become benefits too easily in that approach, and claims then become fragile, ephemeral and hollow. It might be nice that your drug helps patients face a new day, but someone somewhere will want to know how, and how your drug does that better than the one they have.

The next great pharmaceutical positioning *may* be emotional. The best rule of Deep Positioning is to know all the rules, to see which one you can break. But you'd have to be brave or foolish to start there, and the risk is that no-one will know which you are until the product fails to make it to market or flatlines when it does.

Summary

- While emotional components can be included in a product's communications, positioning based on emotions is not effective in differentiating a product.
- There are fewer emotions than there are products, and that choosing an "emotional" positioning is choosing a generic positioning.
- Agencies positioning products based on "optimism," "trust," and "joy" have to ignore that these are generic and have been used thousands of times before.
- Positioning based on negative emotions can be effective
- Emotions are great for finding insights, but should not be used to position a drug; rational positioning can inform emotional messaging.

How you position drugs every day: The Covid case study

There are a couple of reactions to 'positioning' within pharma: one is that it's something Marketing will do later (and that's OK), and the other is that it's not really clear what positioning 'is'...

So, let's take the example of the past couple of years, that you've all experienced. Think about how *you* positioned Covid vaccines, because there's no doubt you did.

When you're explaining them to your daughter, or your grandfather, or your Facebook friends, let's examine your use of *framing* and *value*.

Did you import the general framing of 'a vaccine'? That is, were you hoping to borrow attributes from vaccines 'in general'? Polio, flu, travel vaccines, mumps? It should be a careful consideration: *you* may well understand the traditional role of a vaccine way more than your daughter might. It is easy in positioning to assume that you and your audience understand the exact same thing when you use words, or that they understand nothing, but the hard reality lives in the middle – they probably understand them differently.

Remember, there have been good and bad vaccines approved previously, too, and it's hard to selectively import only positive attributes and experiences. There's a wide range of 'efficacy' out there, which is potentially a statistic that pharma knows better than the general public. For example, this chart from pre-approval...

EFFECTIVENESS OF VACCINES
A COMPARISON TO THE COVID-19 VACCINES

VACCINE	VACCINE EFFECTIVENESS
*Pfizer/BioNTech COVID-19:	95%
*Moderna COVID-19:	94%
Flu:	40-80%
Polio:	90-100%
Tetanus:	99%
Measles:	93%
Mumps:	78%
Rubella:	97%
Hepatitis A:	95%
Hepatitis B:	85-100%

*efficacy achieved in clinical trials

ROCHESTER REGIONAL HEALTH

Some of this seems obvious, but you learn when you do positioning right, nothing should be obvious, or taken for granted. As soon as you accept calling it 'a vaccine' or 'get your flu shot' or 'get a jab' or 'you are now vaccinated', you accept the good and the bad. Alternatives could include prophylactic therapeutic, pre-emptive immunisation, etc. When you leave it 'loose', you're inviting your audience to fill in their own thoughts.

Next, what should one *expect* of a vaccine? Is it prevention of infection, prevention of transmission, prevention of illness, prevention of serious illness or death? All of those (in a cascade effect)? That talks to the value proposition. Is one more important than the other? Is your value proposition 'protection', and on which dimension?

How did you address durability? Did you extrapolate from analogues or from your own modelling? How did your expected duration of effect change the story?

When you said that the vaccine was effective, in what way did you provide evidence, or reasons to believe your claim? Was it statistical significance, or clinical significance? Did you show charts, or data tables?

How did you frame the context: approved vaccine versus emergency use? In this scenario you imported trust in the approval bodies and governing bodies. Who did the talking? Was it you or a chosen expert? Did you provide full transparency or choose your news carefully?

You needed to think through the value proposition to the population, but also to the individual. Were you vaccinating your daughter for her benefit (which benefit did you choose?) or for someone else's?

Think about your use of segmentation. Does your value proposition apply equally to everyone, or is it particularly focused on elderly, or vulnerable patients for example? Clearly here your choice of value proposition was interdependent with your segmentation.

Did you then choose to frame all of the vaccines together, or to pull them apart? Were they a 'class' (for example, 'all approved vaccines') or did you suggest, for example, that the mRNA vaccines were different, and in what way? Did you try to suggest that the mRNA vaccines weren't 'a class', but two very different products (and how did you do on that?)?

How did you explain the Mechanism of Disease? Did you stop at 'virus' or delve into spike proteins, upper vs lower respiratory binding? How did that link to your Mechanism of Effect, or Mechanism of Action? How about your differentiation story? Did your choice of mechanism narrative talk to your immunogenicity, without addressing the downsides? How did that impact your side effect/adverse event story? How much did you need to educate your audience on T cells, spike proteins, mRNA or more, before your story would gain traction?

These, and a hundred other considerations, all attach to the value proposition (which, if you do it right, is interchangeable with the positioning). Now, consider that you're not doing it to just one person – you're doing it to everyone. Does your messaging change depending on who you're talking to? With positioning, your core positioning can't: even though you can vary messaging by audience, it has to hang from your core positioning.

If your proposition is that all vaccines are not the same, that yours is more effective at preventing severe disease in vulnerable populations, can you double check that *that* is what you studied, and how you delivered your headline messages? Or, did you get drawn into standard studies, or meaningless comparisons of headline numbers, framed by someone else's positioning, even though your study populations were different?

Positioning is hard, but we all do it, intentionally or not. Even with this example, you can see how positioning should track into Development choices, and therefore is not cosmetic, or gift-wrapping data someone else chose to collect. Like any other discipline in early phase, how well you do it will have a great impact later.

Summary

- This chapter discusses the concept of "positioning" in the context of pharmaceuticals, specifically the positioning of Covid vaccines.
- It suggests that there are different reactions to positioning within the pharmaceutical industry, with some seeing it as something that Marketing will handle later, and others not understanding what positioning means.
- The chapter encourages readers to think about how they personally positioned Covid vaccines to different audiences (e.g. family, friends) and to examine the use of framing and value in these explanations.
- It's important to consider the traditional role of a vaccine and to be careful not to assume that the audience understands the same thing as the speaker when using words.
- Think about the value proposition of the vaccine, the durability of the vaccine, and the context in which the vaccine is presented.

o-positioning

If you have a 'scientific communications platform' but you don't have a positioning, you have a positioning.

If you have a brand name, or a brand colour, but you don't have a positioning, you have a positioning.

It is entirely possible that you have a positioning, but that it isn't very good. The presence or absence of 'a positioning' is not the key question – it is whether it is good.

In the case of the now ubiquitous 'scientific communications platform', there is a view that it is OK to start saying 'stuff', or even to have animated videos of your drug doing 'stuff', because it is scientific, and therefore must be true, and couldn't possibly limit anyone's perception of your drug. Unfortunately, on the way to deriving the message hierarchy, or even a list of 'key' messages, someone has to decide what 'key' means.

Why would a message be 'key', or a headline? It is because it is central to a narrative. So, two scenarios are possible: you will set a scientific narrative running, and then later you will have a new 'positioning'-based narrative, and the two will co-exist; or, you will set your scientific narrative running and then change it to a new one when the positioning is set. In the first, your competing narratives will compete for attention, and distract both internally and externally. In the second, you will be trying to change or reverse a perception that you yourself created.

On the way to developing a scientific communications platform, you will do a positioning exercise. It might not be called that, but it will be a way that you will set the goal of the narrative, against which you will judge which mechanism of disease to describe, which mechanism of action to highlight, and to set the platform from which to communicate. Unfortunately, a 'platform from which to communicate' is a positioning. If it is done simply to establish a publication plan, there is a good chance that it will be underdone - less to persuade than to generate more papers. The only place message should precede mission is in a dictionary. A communication strategy should follow the overall product strategy, positioning forming the platform from which to gauge the individual needs of the drug.

[An aside here: there has been a terrible idea spread across the industry that positioning can be 'premise, promise, proof', or some version of that. It is not just a wrongheaded idea, but a dangerous one. Positioning does not make a promise. It does not assert. Suggesting that you have 'proof' for your 'promise' turns positioning into a statement of fact today, instead of a way to help your customer see your value, over time. It diminishes the value of positioning to a repackaging of data that have already been collected: a passive positioning at best, a facile narrative at worst.]

On the way to developing a brand name, you will do a positioning exercise. It might not seem that way, as they might just ask you whether you want your brand to suggest 'power' or 'freedom', 'independence' or 'calm'. They will ask you to do that so that you can then score the names you see against those fundamental ideas. Unfortunately, the fundamental idea of your product is the basis of its positioning. If it is done simply as a stepping-stone to a brand-name, there is a good chance that it will be underdone - as something that doesn't matter too much, because the legal part is where the hard work comes in. So, for example, a brand like Otezla or Xeljanz may more easily pass legal checks to see if someone else has registered it, but they are meant to be blank canvas names. Brands like Neulasta or Januvia will have chosen positionings based on rejuvenation or new life, which will be relatively generic ideas, desirable for most drugs. They position their brands, whatever the more specific positioning under the hood.

Many positionings end up being the result either of something that was chosen earlier without realising it (indication selection is another classic pseudo-positioning exercise), or they're campaign focused. There is even a risk that your market research will limit your positioning, if you only look in one place with one customer group. All the research in the world on depression could never have yielded the opportunity in pain that Cymbalta saw.

These throwaway positionings are the positionings that are not 'good', even though they live on in PowerPoint slides. Instead, positioning done properly forms a core architecture for everything that comes later - not just in communication (why do we want to shape a market, what do they need to know that is disproportionately helpful to us?), but in clinical development, regulatory (label claim wording) and more. It is a strategic activity, not a tactical one. And, like all strategic activities with downstream impact, the whole downstream needs to be considered, not just the next phase of a project.

Summary

- This chapter discusses the concept of "pseudo-positioning," or the idea that a company may think they don't have a positioning, but in reality, they do.
- The presence or absence of a positioning is not the key question, but rather whether it is a good positioning or not.
- In the case of scientific communication platforms, it is important to decide what "key" messages mean before developing a message hierarchy.
- Positioning cannot be "premise, promise, proof," as it turns positioning into a statement of fact and diminishes its value as a way to help customers see a product's value over time.
- Positioning may be chosen earlier in a product's development without the team realising it, and may be a result of legal or branding considerations rather than a strategic decision.

"Best"

If 'sorry' is the hardest word, 'best' is one of the most abused.

One thing that will always be critical in positioning is the use and abuse of important words.

One that is easy to use, but wide open to misunderstanding, is the word 'best'... It is as abused as the words 'efficacy' and 'safety' (I promise you - if either of those words is in your positioning, it is a poor positioning...)

Consider this non-pharma question: which country's trains are most punctual, or run most on time? Seems an easy question, with an easy answer? Usually it's Switzerland or Japan that's quoted.

However, consider that 'punctuality' is easy to achieve if you set low expectations. Or allow a lot of 'dwell time' in each station so you can set off at the right time regardless of when you arrive... For example, think about what matters when you say your trains run 'on time'...

The Swiss railway claims to be the most punctual in Europe. It says 89.7% of its trains were "on time" between January and October. Great Britain does not have the same reputation. But during the past 12 months, 88.3% of its rail services were "on time". Does that mean British trains are almost as punctual as the Swiss?

Not necessarily. First, the two countries have different ideas about when a train is "on time". For the Swiss, a train is late if it arrives at the station more than three minutes after the advertised time. In Great Britain, it can be up to five minutes late and still count as "on time" (or up to 10 minutes if it's a longer journey). When you look at how many British trains were no later than three minutes, the figure is 83.7%. Secondly, the two countries aren't measuring the same kind of punctuality. In Great Britain, the regulator looks at when the train arrived at its final destination.

In Switzerland, they don't monitor the trains, they look at the punctuality of individuals - how late did each passenger arrive at whichever station they wanted to get off at? It seems almost as though no two countries have the same standards on punctuality. A US train is allowed 10 minutes' leeway for journeys up to 250 miles - but that increases to 30 minutes for journeys of more than 550 miles. In Ireland and Northern Ireland, a train is "on time" if it's less than 10 minutes late (five minutes for Dublin's Dart network). Australia's rail companies each have their own definitions of punctuality. In Victoria, trains have between five and 11 minutes' leeway, while Queensland's trains have either four or six minutes, depending on the route. Meanwhile, Sydney Trains measures punctuality during only peak periods. In Germany, there are two kinds of "on time". So far this year, 94.2% of trains have reached their final destination within six minutes of the scheduled time, and 98.9% within 16 minutes.

"Comparing the numbers is close to an impossible task," says Ben Condry, associate director of the Railway and Transport Strategy Centre at Imperial College London. "Even if you consider only the percentage of trains on time, there are some very significant differences in measurements and definitions. "Are cancelled trains counted as 'late' or excluded from the data altogether? What about trains which are partly cancelled? What if a replacement bus is provided?" "Are some types of delay excluded from the data - such as extreme weather, ill passengers, strikes?" "When you are late, how late are you?" "Leaving early might be problematic - you leave people stranded - but is arriving early problematic? It's not."

So, layer in 'punctuality' and all of the other things that might matter (comfort, predictability, safety, and so much more) to a train passenger... How easy would it be to derive a simple understanding of the word '*best*' as it applies to a train service? It would be hard. The same, of course, applies to airlines - where the takeoff window and the flight time are frequently manipulated to ensure 'on-time' service, or the calculation from wheels down, rather than arrival at the gate... British Airways used to say it was the 'world's favourite airline' based solely on passenger numbers, even though many other definitions of 'favourite' could be considered.

Yet again, we arrive at a place where measuring what matters is critical, but who decides on *what* matters is largely opaque.

In pharma, there are two things that happen: one is that the poor definition of 'best' (reduction in exacerbation count, for example, or 'pain free' (which can be easily achieved, but still leave a patient in pain) is the one that everyone optimises their studies for. So, we end up with lots of drugs claiming similar data, even though their methods of achieving those numbers may vary widely. The other is that the thing that matters may not even be measured. So, we get claims of 'best in class' when the parameter measured is irrelevant to most.

There's no easy answer to this. Other than a request that people stop using 'best' as if it is somehow self-evident. Ask, the next time you hear it, what the product is best *at*... And, realise that this is where the smart folk go to dig into how to position their drug.

Every word, and every idea, in a positioning is critical. If you wanted to be the world's 'favourite' airline, would your focus be on passenger numbers? If that was your positioning, your team could go in a hundred different directions, and still claim to have been aligned. The more plastic the words you use, in terms of interpretation, the less it is a positioning. This is why taglines that masquerade as positionings are so dangerous - they use words designed to appeal, rather than to communicate with clarity. The task of the 'best' positionings is to communicate with clarity, internally.

Summary

- The word 'best' is often abused and misunderstood in positioning.
- The example given is of train punctuality, with Switzerland and Japan often cited as having the most punctual trains.
- However, punctuality can be achieved by setting low expectations or allowing for extra time at stations.
- Different countries have different definitions and standards for what counts as a punctual train.
- Comparing 'punctuality' between countries is difficult due to varying measurements and definitions. This is true of 'efficacy' too.

Talking a different language

Can you imagine how you'd be perceived if you landed in a foreign country, and just started talking to the locals about what they were doing wrong, in your language? Like a modern day missionary, but with no attempt to use words they understand?

What you just imagined is what most sales reps are forced to do: to land in a physician's office, armed with their book of truths (the detail aid), using words like 'endpoint', 'OS', 'significance on ADAS-Cog' that no physician uses in practice. They may well believe they're saving the physician, but it's possible that no connection is made and everyone goes home bemused.

Most of us know the difference between assertion and persuasion, as ways of communicating. *"If I assert that my drug is best in class, you're at liberty to believe me or not. If I persuade you that my drug is best in class, I've done the work I came to do."*

Messages that follow a passive positioning typically fall into this category, wrapping a feature set into a too-obvious benefit, that the physician could fill in for themselves (that is how passive positionings are derived, of course). It is assertion of a feature, assertion of a benefit, and the hope is that it is believed. But, if you chose the wrong feature, knowing nothing of your foreign land's challenges, you go home with no ally.

Messages that follow an active positioning have an opportunity to persuade. To have arrived at that positioning, some understanding of challenges must have been completed, some non-obvious framing of the value taken place, and messages chosen to be part of that story. The goal of active positioning is to change behaviour, not just to *tell* people to change behaviour.

To do that requires an understanding of their language. Telling clinicians about HAM-D, MADRS or PFS, instantly marks you out as someone who didn't take the time to listen - those are not words or ideas anyone uses in practice with individual patients. It may seem minor, but it shows your foreign-ness to your audience. Not just that you don't understand their world, but that you didn't care to find out.

Summary

- Sales reps often use industry jargon when talking to physicians, which can hinder communication and understanding.
- The difference between assertion and persuasion in communication is important for making a connection with the audience.
- Passive positioning, which relies on the audience filling in the gaps, is less effective than active positioning, which requires understanding and framing the value in a way that resonates with the audience.
- Understanding the language and challenges of the audience is crucial for effectively communicating and changing behaviour.
- Failing to understand and use the language of the audience can signal a lack of interest and understanding in their perspective.

What 3 words?

Get there quickly.

The app What3Words ticks the 'wow' box the way that Shazam does - the 'I didn't imagine you could do that' box... To be able to pinpoint exactly where you are on the planet, using three word combinations, makes a difference not just in places that have relatively explicit street addresses, but those that don't (Tokyo, for example). It's not just the geo-location that's interesting, but the fact that there are enough words to do it...

Over the course of 30 years of positioning, I have found that the best have a two or three word 'summary' that do the same

- pinpointing one product only. The Anti-angiogenic, The Weekender, Depression Hurts, Lower Is Better, Erythroid Maturation Agent, Direct and Expect. There may have been a statement underneath each, but the statement didn't make it better - it just provided more narrative to why.

It's easy to think about why that is.

Let's remember: everyone involved in the prescribing decision is human. There are no machines; everyone involved in the guidelines, in the formulary listing, in the treatment team, in the ultimate decision to prescribe, to take, and to keep on taking, the medicine is a human, a real life person. Their motivations, their values, their consequences, are human ones. While the brain might be a remarkable computer, one of the things it does best is to look for patterns, to simplify, to stereotype. So, it is constantly looking for shortcuts - easy ways to think about complex situations. That is where positioning comes in.

Positioning addresses motivations. Its literal task is to help drive preference at the moment of decision. The 'place in the mind of your customer' is the part that deals with motivation - the amygdala, perhaps. I wrote in the Share of ear/share of voice chapter about my six As model of communication - many of the parts of that chain deal with the internalisation and archiving of ideas.

While even Daniel Kahneman was careful not to say that System 1 and System 2 thinking were binary, or belong to parts of the brain, it is typical that data *support* a decision rather than making it clear and obvious, and that decisions are typically made quickly and *then* backed up by data (if necessary).

So, for storage and retrieval of ideas, it is helpful that they are both potent and easy to retrieve (the electric vehicle company, the safest car, Internet search). No-one will remember all of your data. You are lucky if they remember you at all. They might remember the idea of your drug, if you gave them one. Unfortunately, many companies think that throwing a lot at their customers (Look at these charts! Look at these tables! Look at this detail aid! Look! Look!) will do the job of giving them an idea - *if we give them everything, they can build their own idea...* (To be fair, this is the way that a lot of market research companies think you build a positioning, too.)

If you think that this is an odd exercise to do - to summarise a complex product in a simple phrase, just consider for how long movies have done this... "You'll never go in the water again." "He may be dead but he's the life of the party." "In space no one can hear you scream"

Positioning recognises there is a battle for that amygdala, a zero-sum elbows-out competition to be at the front of each patient decision.

Deep Positioning that precedes data will help your customer frame and archive those data. The handy retrieval device that they use will be the one you chose, if you've done it right. Just making your data sound nice, then, can't be the task - it must also be a ready reckoner for what it stands for, and why they should care. It should also address their very human motivations - make your great data relevant to their human day to day.

Summary

- The app What3Words allows users to pinpoint their location on the planet using three-word combinations, making it useful in places without explicit street addresses.
- The best products have a two or three word 'summary' that can pinpoint the product.
- Positioning addresses motivations and helps drive preference at the moment of decision, by addressing the 'place in the mind of the customer' which deals with motivation.
- Positioning is helpful for storage and retrieval of ideas as it makes them both potent and easy to retrieve.
- Positioning addresses human motivations and helps customers frame and archive data.

The notion of your potion

The role of categories

When you think of 'the cloud', what do you think of?

You probably don't immediately think of this…

Luleå, Sweden

This is Facebook's server farm, 70 miles from the Arctic Circle. We all know the reality of massive server farms, and that "***The cloud***" refers to servers that are accessed over the Internet, and the software and databases that run on those servers, but we prefer not to think about all of it.

When we share to 'the cloud' we're sharing to an idea. We don't really know how any of this works, or where it is, but we're happy that it is 'a thing'. That is positioning - it simplifies and crystallises a complex solution. You understand the 'idea' of the cloud, and that's enough - you know what it is for.

The same can be true of positioning pharmaceuticals. Agencies love to focus on taglines, but that's because they're ad agencies. You certainly have an idea in mind when you think of ad agencies. However, when asked 'what's the idea of your drug?', many will struggle.

An example: is your rheumatoid arthritis drug an anti-inflammatory (that's one idea), an auto-immune drug (another, different idea), a symptomatic agent, etc? Is your statin 'an LDL lowerer' or an 'anti-atherosclerotic'. At some level, it helps you understand extremely complex biology: if you think of your acetaminophen as an antipyretic or a painkiller, those are two ideas of the same drug,

Ideas help you to see something, frame something in your mind. It's the origin of the word.

Definitions from Oxford Languages · Learn more

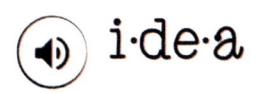

i·de·a

/ī'dēə/

Origin

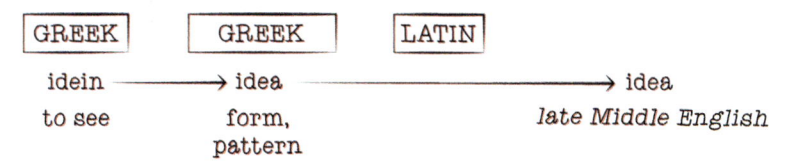

GREEK	GREEK	LATIN
idein →	idea →	idea
to see	form, pattern	late Middle English

A properly formed idea has edges - a form.

How does this help when you're thinking about Deep Positioning? Well, from the earliest moment, it makes sense to play with different ideas of your drug. You may be comfortable talking about your JAK inhibitor as a JAK inhibitor, but the idea of your drug shouldn't be 'it inhibits JAK, here's the data'. Instead, 'immune overactivity reducer' or 'disease modifier' or 'hot joint drug' would be different ideas of the same drug. Each of those ideas might have different development and communication strategies.

There's a key difference between an anti-angiogenic and a VEGF inhibitor, in terms of how you understand its effect: how it works, that it works in combination, etc. It's easy to understand what an anti-angiogenic would bring to cancer; much harder, unless you're down in the weeds of the science, to know where VEGF sits in the dynamics of cancer.

The idea of your drug is something you should be considering from phase I - understanding the attractiveness and feasibility of the different framings. If you are working on a 'negative symptoms' drug, could it be about 'enabling sociability' or 'helping someone get back to work' or a different definition of 'efficacy' in the disease? All of those would be better than thinking you're just developing an 'anti-schizophrenic', an idea that contains every other drug in the space. Perhaps your 'anti-schizophrenic' is actually an 'anti-depressant', a different idea? Your physicians may well understand negative symptoms, but they may not understand how your effect size might translate into value unless you help them. Perhaps your drug could be the only one to target rumination, a more specific, ownable idea?

The 'idea' of your drug is a simple framing tool - but achieving simplicity is not easy. It forces you to spend time thinking of the value of your drug, and it's never too early to do that.

Summary

- This chapter discusses the concept of "positioning" in relation to server farms and pharmaceuticals, and how it simplifies and crystallizes complex solutions.
- The idea of a drug is an important aspect of positioning and should be considered from the earliest stages of development.
- Different ideas of a drug may have different development and communication strategies.
- It is important to understand the attractiveness and feasibility of different framings of a drug.
- The idea of a drug is a simple framing tool that forces you to spend time thinking about the value of your drug, and it's never too early to do that.

Time flies like an arrow; fruit flies like a banana

The role of analogy

Analogy is at the core of positioning. Because you are seeking to provide a frame through which customers might experience your product, it is important to consider how you might wield the concept.

In the popular phrase 'Time flies like an arrow; fruit flies like a banana', you see both analogy and assertion. We all know what we mean when we say 'time flies like an arrow', even if it takes some interpretation - we're working with the *concept* of time, and our mental concept of an arrow. In 'fruit flies like a banana', we have a more obvious statement of data.

In 'like an arrow', no-one thinks we mean its shape, because we said 'flies like'. But we also take the idea of fast and straight, instead of 'only for 30 metres'. If we were to say time is money (or 'time is muscle' as in a potent positioning for 'clot-busters'), Time is a flowing river, or time is a gift, we don't have to argue about which analogy is best - we can get the concept on both sides of the analogy. We don't have to invite the regulators to a conversation about whether 'time flies *like an arrow*' or 'time flies *when you're having fun.*' The word 'flies' can have many meanings - useful if that's what you want, but also a caution for those ambiguous words in positioning statements.

In positioning, there are only two ways to leverage this insight. Your product, or your company, offers either:

- Sameness with difference (quantitative differentiation)
- Something different (qualitative differentiation)

'Sameness' relies on analogy. It says 'that thing you understand - we're like that, but better in this key way'. You get to define the category. Whether it is 'a TKI but better in this way', or 'anti-inflammatory, but with more impact on disease activity', you are relying on people's framing of the first concept to provide your position.

If you follow the qualitative differentiation route, you don't have that building block to start with. You are going to have to provide your own category. But therein can lie real opportunity.

Often, if positioning is not working, it is a category error that sits below. Typically, you might hear 'despite having the best efficacy, physicians are saving it for later line.' The words 'efficacy' and 'best' are rarely unpacked.

There are only two reasons someone might have a different opinion than you do:

- You know something they don't
- They know something you don't

(It is frustrating how often people assume it is always the first of these, in life as well as in pharmaceuticals.)

If physicians are using cheaper, less 'effective' products than yours, it could well be that they are unaware of your proposition, and it is awareness that you need to build. However, they are balancing a lot of things, and their reading of the situation may provide a category framing where your product is just an expensive analogue of what they're getting done a different way - their trade-offs might be known to them, but unknown to you.

For example, a cheap anti-inflammatory might have the value of familiarity, rapid onset, easy to prescribe and justify to the healthcare system, and more. If you have asserted that the value you would provide in that situation should outweigh those considerations, they may choose to have a different opinion than you. And that might be because you failed to listen to the things they know that you don't, and then pursued a 'sameness with difference' positioning - and they then tell you that the difference is not enough to overcome the incumbents.

'A place in the mind of your customer' relies on analogy - to 'place', to the concept of 'mind', to the concept of 'customer' (which in pharma, almost more than any other industry, is a complex concept). Analogy is a potent tool in positioning - but too much of it looks like the banal assertion of flies liking bananas.

Summary

- Analogy is important in positioning as it helps customers understand a product or company
- The phrase "Time flies like an arrow; fruit flies like a banana" illustrates the use of analogy and assertion in language
- In positioning, there are two ways to leverage analogy: by offering sameness with difference or something different
- Sameness relies on analogy and requires the company to define the category their product belongs to
- If positioning is not working, it may be due to a category error or failure to understand the customer's perspective and trade-offs.

Knowing

Getting sign off

Deep positioning in early phase means generating market position options. There is a rational decision science that will help teams decide between those options. The 'sign off' for these options should be a process of deciding how best to study them - the 'next best experiment' test.

Positioning in later phase typically means someone senior will want to sign off. Unfortunately, rather than trust the team process, typically this will involve a short review, and a challenge during which all of the myths can be brought out: 'it needs to be more emotional; has it been tested?; shouldn't we wait until this campaign runs before changing everything?; we can't say that…'

If the process to derive the positioning was robust, matrixed and effective, there is no-one on the planet who can propose a better alternative following a short slide review.

It is possible to identify a weak construction, or an absence of logic, or that the statement is a simple string of 'nice' sounding phrases, and suspect there must be something better out there, but there is no easy fix at that point. Typically it ends in wordsmithing. If this positioning was the best of a weak bunch of choices, the issue is with the process, not with the choice.

Easy objections, such as 'it needs to be more emotional' might sound like the voice of experience, but more often reveal inexperience, or lack of confidence, in positioning. Typically, such companies' positionings all look the same, changing just the product name. Even more typically, the company will repeat the process regularly, seeking to rectify a failing campaign. A different agency, a different brand lead, and yet the outcome will be the same, limited by the inexperience of the senior leadership.

Having seen hundreds of positioning 'statements', the product of weak process – and the same process mistakes are repeated across most can be quickly spotted. But on the spot, it's almost impossible to suggest a better one as no-one yet knows what the team knows. It is easy to resort to wordsmithing - we all become poets and authors and ad executives instantly.

There is only one solution to senior management sign-off for positioning: they must sign off on the process, not the solution. A leadership that has empowered its team should send them off on the journey, and not then try to helicopter into the end stages with personal preferences, ignoring the significant interactions the team will have had on the way. Great process will reveal things about the product that the team didn't know before, or about the market. Missing those insights means that anyone outside the team is relying either on what they knew before, or hoping it can all be crunched into a few minutes of briefing.

We can admit that there are ways that process can be corrupted: if the decision making within the team is *too* democratic, or there is too strong a veto from key departments (regulatory, market research, perhaps), the voices of experience in the team can spend their time on team wrangling rather than choosing bolder, better options. There is no surprise that the best positionings I have ever seen, or been involved with, were generated by small teams of experienced executives, empowered by leadership. 'Buy in' can be overrated. Teams certainly need to understand the ins and outs of the chosen positioning. They don't all have to 'like' it, but they do have to agree to implement it. A strong positioning architecture is specific and detailed enough to direct activity within guardrails, and so is not amenable to a popularity contest.

There is also the 'Miss America' challenge. The "Miss America" decision-making error refers to a phenomenon where people vote for a candidate they think will win, rather than the candidate they personally believe is the best choice. This is also known as the "bandwagon effect" or "herding behaviour." The phenomenon is often observed in situations where people are unsure of their own opinions and look to others to guide their decision-making. In positioning process, this shows up as people choosing what they think is the best 'positioning' rather than bringing their perspectives to the strategy. Positioning has to do work, rather than sound 'nice', but on first glance, many people want to go to taglines rather than deep positioning.

If, instead, senior management want to be involved, they should either be part of the process, or let the team make their own choice and support them. If leadership don't want to be part of the process, they have to choose the latter option. If they don't, they are effectively saying that they don't trust the process. If that is the case, it is a waste of time and resource to let the team start the process in the first place.

The best leadership establishes a corporate process that seeks Deep Positioning in early stage, and active positioning in later stages. It then allows the teams to run the process without second guessing and reviewing the outputs. Trusting the team is binary: if they are allowed to run the process, they must be allowed to own the output. Removing that empowerment means removing the trust.

Summary

- The process of deep positioning in the early phase involves generating market position options and making a rational decision through the "next best experiment" test.
- Positioning in the later phase often involves senior sign-off, which can lead to objections and wordsmithing instead of a solid position.
- To prevent this, senior management should sign off on the process, not the solution, and empower the team to carry out the journey.
- The "Miss America" challenge refers to people choosing what they think is the best positioning instead of bringing their own perspectives.
- The best leadership establishes a robust positioning process, trusts the team to run it, and allows them to own the output.

Testing

When it comes to 'testing' positioning, the first question that should always be asked is 'why?' Knowing whether a positioning will work is one question. Knowing whether it is the *best* positioning is a different question, although they are often compressed into one 'research-like' process. Typically, there are different roles for 'testing' or research in positioning exercises: the search for unmet need against which to position (usually called 'insights'), the evaluation of positioning options, and the improvement of a final option. This chapter will mostly focus on evaluation.

First, let's establish this fundamental point: very few of us experience positioning in our day to day lives. Done well, it is only an internal construct that turns into things people experience. So, who are the people who can tell a good one from a bad one? Oddly, traditional testing takes this internal-only component and road tests it.

The way a product is presented or "framed" can greatly impact its perception. This is particularly true in the case of a new drug, where the framing can determine whether it is seen as a first-line treatment option or a last resort. The importance of a successful positioning is critical and should be approached actively, rather than passively.

As covered earlier in the book, one approach to positioning is the use of passive positioning. This is when data is simply presented and the audience is left to make their own conclusions about the product. However, this approach *cannot* then be effective in shaping the audience's perception of the product – it is a circular proposition, defining a product only as what is obvious and by what has come before - the audience's *existing* perceptions. Instead, an active approach to framing is necessary.

Active framing involves making deliberate decisions about how to present the product in a way that influences the audience's understanding. This can involve highlighting specific features, comparing the product to others in the market, or presenting it in a specific context.

Market research is often used in the positioning of new drugs, but the traditional approach can lead to stereotyping and a lack of flexibility in the perception of the product. Specialists chosen for market research may have a preconceived notion of the product's place in the market, limiting the ability to shape the audience's understanding. They will typically have deep understanding of one field. What if that is the wrong field for your drug?

For example, describing a non-steroidal anti-inflammatory drug as a potential cure for Alzheimer's may initially confuse some physicians. The traditional perception of NSAIDs as 'anti-inflammatory' would limit their imagination and understanding of how it could affect a neurodegenerative condition. Perhaps their experience would tell them that other NSAIDs don't. As soon as you frame the story, there is a limit to how far the physician will travel in 'her imagination'.

On the other hand, presenting the ideas behind the product first ('neuro-inflammation as a target, perhaps), rather than its category and its TPP, allows the audience to use their own imagination and projection, and provide a unique and compelling positioning.

The problem with passive positioning is that it's difficult to shift people's perceptions once they have been framed in a certain way. Traditional market research ignores that truth, and tends to rely upon passive positioning and quantitative differentiation as the core of its methodology, which can result in false positive views of a product's effectiveness. For example, imagine up-front in research showing a TPP that says a drug is 10% better than a current product. Presented with that 'cold' information, physicians may only then consider it for use in certain situations where incremental added benefit is sought. Is that 10% on a linear scale, does it provide an absolute benefit that couldn't be achieved any other way, can physicians even tell the difference?

For example, not all improvements are perceived in a linear fashion in the real world. Kano's model is a great illustration of that. A popular tool for understanding customer needs and preferences, it can help organizations prioritize customer needs, evaluate the effectiveness of their product or service offerings, and develop strategies to better meet customer expectations.

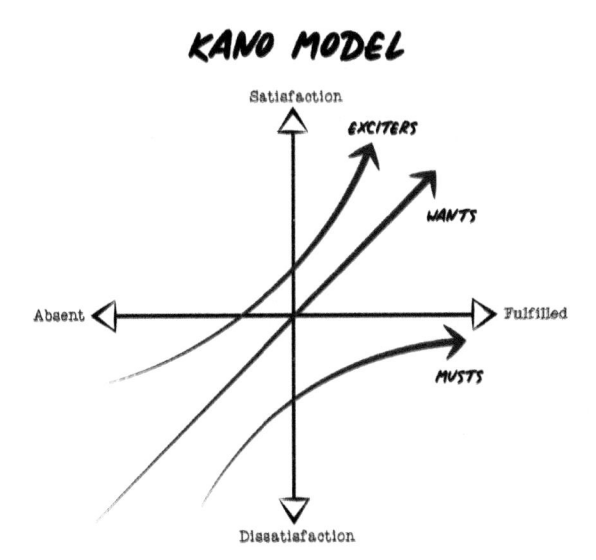

KANO MODEL

Kano's Model is a customer satisfaction theory that categorizes customer preferences into five types of attributes: Must-Have, Performance, Delighters, Indifferent, and Reverse. These attributes are used to analyse and prioritize customer requirements, and to guide product development.

- Must-Have attributes are the basic necessities that customers expect and take for granted, e.g. reliability of a product.
- Performance attributes are the aspects of a product that directly affect its quality, e.g. speed.
- Delighters are the attributes that exceed customer expectations and can provide a significant source of differentiation and customer satisfaction, e.g. style.
- Indifferent attributes have no effect on customer satisfaction, e.g. colour of a product.

Reverse attributes are the aspects of a product that, when present, negatively affect customer satisfaction, e.g. complexity.

At its core, Kano's model proposes that customers' satisfaction with products or services is determined by three factors: basic needs, performance needs, and excitement needs. Basic needs are the foundational requirements of a product or service; they must be met in order to satisfy customers at the most basic level. Performance needs are those features that customers expect from the product or service; these features increase satisfaction as they are improved upon. Finally, excitement needs refer to features that delight customers; they provide an extra motivation to purchase and use a product or service. Now, imagine presenting a TPP up front in testing, and then asking physicians how they'd 'position' the drug. At its core, it is a collection of features, presented 'cold' - even a quick review of Kano's model shows that a 10% change on the raw axes will make a different impact, depending on the feature being a must, a want or an exciter.

For example, you already know that a car that can do 35mpg is not as good as a car that can do 40mpg. Is the new option 14% better? Pharma would typically present it that way. What if this new car came with 5,000 mile service intervals instead of the 10,000 of the older one? The way most pharmaceutical positioning testing works is by exploring the top-line mpg value to the respondent.

A lot of this research error is fed into the *generation* of positioning options. Unfortunately, the way that positioning is 'tested' can compound this mistake.

First of all, the *idea* of testing positioning is problematic. What exactly is being tested? As Deep Positioning is a strategic choice, it is important not to ask if the positioning being tested is a good 'positioning'. Many research processes simply try to establish the likeability of a statement, or ask the physician to rewrite it to 'sound' better. As I said in the previous chapter, it is impossible to know if a positioning is good on first impressions. Testing in this fashion tends to prefer nice-sounding statements, especially ones that are vague enough to be widely interpreted.

Secondly, the choice of respondent is critical. Physicians know how things are today, and how things used to be. They will rarely know how the upcoming competitive landscape will look, especially if they have a narrow speciality (or are too broad – a typical challenge in finding research respondents). The likelihood that they are representative of the *future* audience is low. The likelihood that they will be able to understand their own motivations and decision process is also low, so a positively-presented statement will tend to do better. These two factors mean that testing positioning can lead to false positive views of its effectiveness.

So, why test at all? It might seem odd, but its main role is to give people who don't understand positioning some confidence in the final positioning.

Typically, it is linked to the need for senior sign-off. Of course, that relies on two things: showing it is the 'best performing' option, and showing that it is likely to positively impact perception. This 'false positive' problem becomes an issue in this context. Nice-sounding 'positionings' do better in 'testing' and so tend to attract more easy sign-off, despite having no real world relevance.

Testing positioning options can be problematic in several ways, including:

1. False negatives: Testing may not detect a positioning problem even when it exists, leading to misdiagnosis or missed opportunities for improvement.
2. False positives: Testing may detect a positioning success that doesn't actually exist, leading to undue confidence. For example, an option being just the best of three poor options does not mean it is a good choice.

3. Limited specificity: Testing may not be able to distinguish between different types of positioning, making it difficult to determine the best course of action. Testing a bold, effective positioning that relies on an understanding of the disease that can only come from market shaping/ education against a 'tagline plus' positioning will inevitably favour the nice-sounding and easy one.
4. Observer bias: The results of positioning testing can be influenced by the person performing the testing, leading to inaccurate results. It is often unclear why a certain agency was chosen, but it is clear that the same outcome would not be achieved were different agencies to be applied - this means that the outcome is not necessarily objectively the best, but the one favoured by the agency chosen.
5. Inter-observer variability: Different people may interpret the results of positioning testing differently, leading to conflicting conclusions. It is rare not to have all observers used the words 'I think...' when detailing their conclusions, which suggests that objectivity is missing.
6. Unreliable results: Positioning testing can be subject to measurement error, leading to unreliable results and difficult-to-interpret data. It is impossible for it to be representative, so it can only be suggestive. One issue remains the averaging of the answers - the 'average' of 50 people hating an idea and 50 people loving it is not 'indifference'.
7. Inconvenient or uncomfortable: Some positioning testing may be invasive or uncomfortable for the person being tested, which can limit the willingness of people to undergo testing. This assumes they were even the right kind of people to be in the testing. It also assumes that they can mentally project themselves into a future in which a whole lot has changed.

One way in which testing can be useful is in looking for objections, or unintended consequences. Words that seem perfectly innocuous in constructing a positioning can mean very different things to different people - it is useful to hear that in research before they're used in the real world. For this to be useful, the research really has to listen, not be looking solely for improvements, or for a winner.

That kind of testing is exemplified in the process of improving prototypes, covered in the chapter on ideation and prototyping - you're looking for the broken parts.

It is hard to find any good defence of testing for positioning. The test of a positioning is whether it makes a difference in the market. Few of the best positionings were 'liked' in testing, although they did score highly for memorability, differentiation and more.

Summary

- The first question when it comes to 'testing' positioning is 'why? What are you testing?'.
- The success of a positioning is a crucial aspect that should be approached actively instead of passively.
- Active positioning involves making deliberate decisions about presenting the product to influence audience understanding.
- Traditional market research relies on passive positioning and quantitative differentiation, which can lead to false positive views of a product's effectiveness.
- The effectiveness of a product's positioning can be analysed and prioritised using Kano's model, which categorizes customer preferences into five types of attributes. The model proposes that customer satisfaction is determined by three factors: basic needs, performance needs, and excitement needs.

Conclusion

In the journey around Deep Positioning, we've examined a lot of 'why', some 'what' and some 'how.' If your main takeaway is that you need to start in Phase I, with a matrixed team of equal voices, and with a range of market positions in mind, rather than one, all of that will have paid off.

Deep Positioning *is* what you do to your product, and then of course what that does for the hungry mind of your customer. Understanding that one thing – that a product is not a molecule – may be the most important idea in this book.

We're not an industry that launches many products per year. Lessons from success are harder to rely upon than the lessons from 'failure'. Drugs do what drugs do – the challenge is to find out where that might be, what might be approvable, and which combination of those two is commercially attractive as an investment.

When I said, in the Prologue, that the great positionings are not the ones that stop at the PowerPoint slide, but in every single thing that drug goes on to do, you can see why Deep Positioning is the more relevant discipline. It is an energising idea, a collaborative idea. It is an idea that needs application, and to do that will take a lot of time to get right. Best start now.

Appendix

Top 30 Pharmaceutical Positionings

Positioning can seem subjective, to the uninformed. But it is not. A good positioning carries out work for its product – objective, measurable work. A great, active positioning transforms a good product. Objective outcomes are all that matter – we know, as we've been lucky enough to work with eight of the 15 biggest launches of the past five years.

However, we didn't work on all of them. So, as we do from time to time, we asked ourselves the question: what are the best positionings of all time in our industry? Using some objective measures, and the collective wisdom of IDEA, we argued and debated, agreed, and then argued some more.

But we did create a list of the 30 best positioned drugs of all time. We didn't work on many of them, but we have studied them, for clues… (We've also studied the 30 worst, but that's for another day…)

We're open for debate, argument and more. But, for now, from the most successful positioning practice in pharma, here's the definitive Top 30:

1. Revlimid (lenalidomide) – Oncology
2. Cymbalta (duloxetine) – Neuropsychiatry
3. Nexium (esomeprazole) – GI
4. Opdivo (nivolumab) – Oncology
5. Ocrevus (ocrelizumab) – CNS
6. Lipitor (atorvastatin) – CVM

7. Avastin (bevacizumab) – Oncology
8. Viagra (sildenafil) – GU / Sexual Dysfunction
9. Adderall (amphetamine, dextroamphetamine) – Neuropsychiatry
10. Cialis (tadalafil) – GU / Sexual Dysfunction
11. Gilenya (fingolimod) – CNS
12. Humira (adalimumab) – Immunology
13. Rituxan (rituximab) - various
14. Victoza (liraglutide) – CVM
15. Advair (fluticasone, salmeterol) – Respiratory
16. Abilify (aripiprazole) – Neuropsychiatry
17. Biktarvy (bictegravir, emtricitabine, tenofovir alafenamide) – Virology / Infectious Disease
18. Enbrel (etanercept) – Immunology
19. Fasenra (benralizumab) – Immunology
20. Spiriva (tiotropium) – Respiratory
21. Xarelto (rivaroxaban) – CVM
22. Xalkori (crizotinib) – Oncology
23. Orencia (abatacept) – Immunology
24. Propulsid (cisapride) – GI
25. Aricept - (donepezil) – Alzheimer's Disease
26. Tamiflu - (Oseltamivir) - Influenza
27. Eylea (aflibercept) – Ophthalmology
28. Losec/Prilosec (omeprazole) – GI
29. Latuda - Anti-Psychotic
30. Premarin (estrogen) – Women's Health